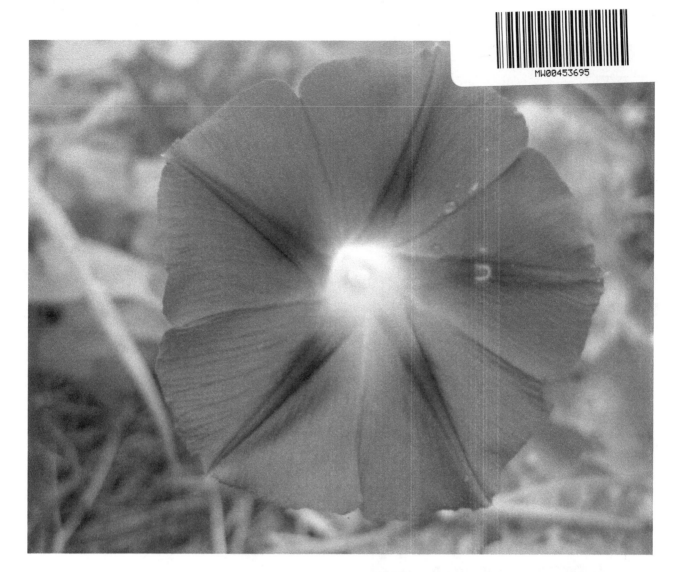

FRIENDLY

BIOLOGY

Joey Hajda DVM MEd

Lisa B. Hajda MEd

Student Workbook

Published by Hideaway Ventures, 79372 Road 443, Broken Bow, NE 68822

For information regarding this publication please contact Joey or Lisa Hajda at the above address or visit our website: www.friendlybiology.com.

Copyright 2017 Hideaway Ventures, Joey or Lisa Hajda

Front cover photo credits: *Tramea onusta.* (Red Saddlebags dragonfly) Tim Hajda; *Papaver spp.* (poppy) Joey Hajda; *Thamnophis spp.* (gartersnake) Tim Hajda and our granddaughter Phoebe Jean by her mother, Clara Williams, Clara Williams Photography.

Table of Contents

Lesson 1..5

Lesson 2..9

Lesson 3..11

Lesson 4..15

Lesson 5..19

Lesson 6..22

Lesson 7..24

Lesson 8..28

Lesson 9..31

Lesson 10..34

Lesson 11..37

Lesson 12..41

Lesson 13..45

Lesson 14..49

Lesson 15..52

Lesson 16..53

Lesson 17..55

Lesson 18..59

Lesson 19..61

Lesson 20..64

Lesson 21..66

Lesson 22..71

Lesson 23..74

Lesson 24..76

Lesson 25..78

Lesson 26..82

Lesson 27..84

Lesson 28..87

OUR PETS

TROTWOOD

BUD

SHIRLEY AND OLIVE

MARTHA, BABY 1, BABY 2, ETHEL AND GEORGE

JEAN LOUISE AND BONNIE

VELVET AND TIM

IF YOU HAVE PETS, DRAW THEM HERE!

Name _____Date_____

Lesson 1   Practice Page 1

Instructions: Fill in the blank with the appropriate word. Refer back to the text portion of the lesson for help.

1. The term biology is derived from two words: _____ which means life and _____ which means the study of.

2. There are other words like biology. The study of the earth is _____ while the study of ancient artifacts is _____.

3. Hematology is the study of _____ and the study of tumors or cancers is known as _____.

4. Being able to _____ or change location or position is evidence of life in a living thing.

5. A feature common to all living things is that they can make new living things. This process is known as _____.

6. Living things all require a source of _____ in order to carry out life processes. For humans this source is the _____ we eat, however for most plants, it is the _____ which provides this source.

7. Small living things increase in size and complexity. Living things _____ and develop.

8. Living things react or _____ to things around them which is also known as their environment.

9. _____ is the condition where evidence of life no longer exists.

10. The name of the primary author of this textbook is _____.

Read each example below and tell whether which form of evidence of life is being demonstrated. Some examples may have more than one evidence of life present.

Example: A puppy chases a ball. Movement

11. Tommy is hungry and so he eats three hot dogs for lunch. _____

12. A tree sparrow flees from an approaching cat. _____

13. The goose in incubating ten eggs. _____

14. The earthworm wriggles down deeper into the bedding in the bait cup.

_____

15. The sunflower tilts its flower towards the east in the morning hours.

_____

16. Mushrooms tend to be found most frequently near dead or dyeing trees.

_____

17. The kitten learns to catch mice by imitating its mother. _____

18. As the summer month go by, the ears of corn on the corn plants appear to be getting larger in diameter. _____

19. Allowing food to sit out unrefrigerated and then eating it can result in food poisoning.

_____

20. The peach tree developed pink blossoms as the temperatures increased.

_____

Choose an animal in your home or one you are familiar with. Tell what it is and then give four pieces of evidence that tells you it is a living thing. For example, if you choose your dog, a piece of acceptable evidence is that it needs to be fed everyday.

21. _____

    1.

    2.

    3.

    4.

Choose a plant in your yard or one you are familiar with. Tell what it is and then give four pieces of evidence that tells you it is a living thing.

22. _____

    1.

    2.

    3.

    4.

Name _____Date_____

Lesson 1   Practice Page 2

Instructions: Below you will find clues to solve this crossword puzzle. Refer back to the text portion of the lesson for help.

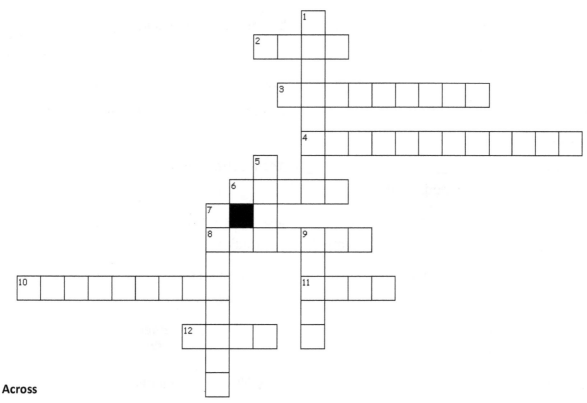

Across

2. living things _____ and develop

3. process of a living thing making another living thing

4. requirement of living things in order to carry out life processes

6. when evidence of life no longer is present in an organism

8. living things _____to their environment

10. author of this textbook

11. tumors or cancer

12. life

Down

1. changing position or location as indicator of being alive

5. earth

7. one complete living thing

9. the study of

Name _____Date_____

Instructions: Fill in the blank with the appropriate word. Refer back to the text portion of the lesson for help.

1. All things, whether living or non-living consist of tiny bits of matter known as _____.

2. The central portion of an atom is known as the _____ and it contains subatomic particles known as the _____ and the _____.

3. Circling around the nucleus of the atom are a third type of subatomic particle which are the

_____.

4. Theories say it is the _____ of the electrons which determines the behavior of various elements on the periodic table.

5. To determine the number of protons or electrons an atom of a particular atom has, one looks for the _____ of that element on a periodic table.

6. The electrons are thought to exist in _____ around the nucleus of an atom and that there can be no more than _____ on one of these layers.

7. Elements which have their outermost layer of electrons filled are the elements which are very _____ in their behavior.

8. Elements which have their outermost layers incompletely filled are elements which are very _____ in their behavior.

9. Elements which are very reactive seek to gain stability by moving or sharing _____ with neighboring atoms of elements.

10. The family of elements whose atoms have their outer layers of electrons completely filled making their very, very stable is the _____ family.

11. Atomic bonds which form between atoms who have transferred electrons from one to another are known as _____ bonds.

12. Atomic bonds which form between atoms who are sharing electrons between themselves are known as _____ bonds.

13. Of the many elements known to man, there are four that are common to all living things and deserve our special attention. Those four elements are: _____, _____, _____ and _____. Write their element symbols next to their names, too!

14. Of the elements listed below, choose the one that would most likely be the least reactive.

    A. Hydrogen

    B. Carbon

    C. Sodium

    D. Neon

15. Of the elements listed below, choose the one that would most likely be the most reactive.

    A. Neon

    B. Sodium

    C. Argon

    D. Helium

16. Which subatomic particle is thought to be responsible for an atom's behavior?

    A. Proton

    B. Neutron

    C. Electron

    D. Crouton

17. Suppose Atom A desires to get rid of one electron and Atom B is willing to accept that one electron. Together, by moving this electron, they can become a compound which is stable. This type of bond formation where electrons are moved is called a

    A. Proton bond

    B. Single covalent bond

    C. Double covalent bond

    D. Ionic bond

    E. James bond

18. There are many elements required by living things in order to maintain life. There are four that we will study next. Circle these four important elements found in this list of element symbols:

H   He   C   Ca   Li   Be   Ar   Ox   O   Br   Ni   N   Zn   As   Pb   Cu

Lesson 2   Practice Page 2

Instructions: Below you will find clues to solve this crossword puzzle. Refer back to the text portion of the lesson for help.

Across

3. subatomic particle analogous to planets in our solar system

6. defined as a change in position or location, a feature of living creatures

10. It is the _____ of electrons which determines reactivity of an element.

11. a feature of living things where more living things just like the parents are created

13. an important element found in living organisms, a component of our air

14. the central part of an atom

Down

1. the type of bond formed when electrons get transferred from one atom to another

2. an important element of living organisms which has the atomic number six

4. tiny bits of matter from which all things are thought to be made

5. the type of bond formed when atoms share electrons between themselves to gain stability

7. an important element of living organisms which has the symbol N

8. an important element in living organisms, also found in water

9. electrons are arranged in _____ around the nucleus of an atom

12. subatomic particles found in the nucleus of atoms

Name _____Date_____

Lesson 3   Practice Page 1

Instructions: Fill in the blank with the appropriate word. Refer back to the text portion of the lesson for help.

1. The common name for carbohydrates is _____.

2. The word carbohydrate is derived from two words carbo– which indicates the presence of _____ and -hydrate which means _____ is also present.

3. The generic formula for a carbohydrate is _____ where the little letter "n" tells _____.

4. The carbohydrate in which n = 6 has the formula _____ and its name is _____. Its common name is _____,

5. Another carbohydrate in which n = 6 and is found in fruits is known as _____. Its structure is slightly different from that of glucose and is therefore called an _____ of glucose.

6. A carbohydrate made from one type of sugar molecule is known as a _____ where the prefix _____ means one and the root word _____ means sugar.

7. Two examples of a monosacharride are _____ and _____.

8. A carbohydrate which consists of two types of sugar molecules is known as a _____. A common example of this type of sugar is _____ commonly known as table sugar.

9. The two monosaccharides which make sucrose are _____ and _____,

10 _____ is the carbohydrate found in dairy products. It is classified as a _____ because it is made up of two simpler sugars. Those two are _____ and _____,

11. The fuel for living things is _____.

12. More complex carbohydrates can be broken down into glucose through the action of _____.

13. Enzymes primarily work in two ways: _____ or _____.

14. The scientific word for cutting is to _____.

15. The enzyme which breaks down table sugar is _____.

16. The enzyme which breaks down milk sugar is _____,

17. People who are deficient in producing enough lactase are said to be

_____,

18. _____ is a disaccharide found in seeds and grains. The enzyme which breaks down this carbohydrate is _____,

19. Plants have the capability of joining several glucose units together into complex structures. These complex structures are known commonly as _____ and scientifically as

_____,

20. The enzyme capable of lysing amylose is _____ and can be found in our

_____,

21. _____ is a carbohydrate that is not sweet tasting, but instead functions to provide strength and support to plants. It is formed by very long chains of glucose molecules bonded together.

22. The enzyme which breaks down cellulose is known as _____.

23. Microorganisms within the stomachs of ruminants and termites work to the benefit of their host species. Tell or draw a picture of how this relationship happens.

24. A type of relationship where one organism gains benefit while the other suffers is known as a _____ relationship. Give two examples of this type of relationship.

     1.

     2.

Name _____Date_____

Lesson 3   Practice Page 2

Instructions: Below you will find clues to solve this word find puzzle. Refer back to the text portion of the lesson for help. Note that the words may read forward, backward, up, down or at a diagonal.

1. Common name for carbohydrate is a _____.

2. Cn(H2O)n is the generic formula for a _____.

3. Carbohydrates are made up of three elements: _____, _____ and
_____.

4. Another name for a carbohydrate is a _____.

5. A carbohydrate made up of only one type of sugar molecule.

6. A carbohydrate made up of two types of sugar molecules, such as sucrose.

7. These compounds can work like matchmakers to speed reactions or scissors to cut apart chemical compounds like carbohydrates.

8. Scientific name for table sugar is _____.

9. _____ is the carbohydrate which fuels living things.

10. This sugar makes the sweetness of fruits.

11. The carbohydrate of dairy products is _____.

12. Plants store carbohydrates in the form of _____ and the scientific name is _____.

13. This type of carbohydrate is not sweet like other carbohydrates. It forms cell walls in plants and gives plants strength and "crunch."

14. This enzyme has its action specifically for table sugar.

15. This enzyme has its action on the sugars found in milk and cheese.

16. People who lack the lactase enzyme are said to suffer from _____.

17. _____ are relationships between living things which may or may not be helpful for each other.

18. In this symbiotic relationship one member benefits while the other suffers.

19. In this type of symbiotic relationship one member benefits while the other is not directly affected.

20. The enzyme which breaks down cellulose is known as _____.

21. _____ like cows, sheep and deer, have a symbiotic relationship with microorganisms to digest cellulose. Insects, such as _____ also have _____ in their digestive tracts to digest cellulose.

22. The chemical formula for glucose is _____.

```
E S A T C A L I K O B A N T A P S M Z A U P V I M T S V X X F U R E I T V Y K S
P M R M V O P H S D N E O K I I W K J E L H Z T Q N K Q B K R F U B C K S U U U
D H B C E T D O U F I T K C S U G H L Z D I K Z P C C D S F L D M U X P D C R W
J H G G Z I P C F Q W V N O K W E P B Q E E Z G M Z M C B E C Z I G W J R N Y A
M S I L A S N E M M O C I E I T D Q Y K V N H M R E E U D C X Z N M Y A H S Z S
H Z O U T T Y E I S Z B D I P E L A A T L E F J C W P E E N K T A B S R O M H Q
W L Z Z X C U J M L M U M A L D Q D R F O F D X J M P F W A C R N E B Z U V M Q
H W S U N N B M S Y J K P U X R O F E T F U W J H C A A T R V K T H X W H C Q P
Q C E L L U L O S E Q J N M I L Y C D L P J A Z J A Z R W E Q P S T G A R E J D
H M D U S B X E T A M W J E A U M D N B L K R Y W R A F F L Q D N F L S B A S X
X M R D V E Y W I I V I A K J N N O S G O J O C X B M H T O D N W U R F D C Y C
P R K D H I D I Y N C E C T I C U Z K G L B K I X O Y Y R T G L U C O S E Q C U
A F Z S U K V I S U G A R R E H W J V H L Y S M S H L O A N F R F J N F T T J N
J O O S W X Z G R Z S N Y J O R S Q W Y L O F T J Y O C Q I M S W E M W C O K G
M O N O S A C C H A R I D E D O S E X H E D V M X D S I B E J C E S O T C U R F
R D T L J I L T O S H W Y I V R R Y T V A R B E L R E K J S D G E G D H U X N E
H H G R X I T P C G O C S L A E W G L I P K N N Q A L G D O U E V L M X X W A K
Z T A U Z O M B S I Q A C N W L O E A Z M W L C D T Q W X T W D L Z L I T R F Z
B L H U T A U Z H E C C T A S O W L O N B R G X S E V M H C A D W L S U M K H G
M H I Z Y W U C H C G P A O S T M N A A I A E R E T B J P A O B C N U P L Y Z J
U L K O H A R O H Q Q G I H H E S Z O C L S F T M J X T N L F F L V M B D A W P
Z H L I Q A Y A N V O E C X S M I K L B T B M M Y G Q M L T D E X Y D R U S S U
A Y W V T Q R M E P B A I O D K T J B J R O W S Z T A K D C O M G E O Q O A T E
E E X S G I M H G P D S R G Y A I C T N S A S H N Z T E N R V V F G W P W H E F
Q I E Y D C A D Y I C C P X R J S V I C F G C E E Q T W M O Q N E Z A H N J C Y
G H C E X V H B X G U F J Q P L A A R J J Q W D K U V G P F Z N R W S M H W M K
D L M H D A K Q O S W U P B W F R H K X A D F S F A K T J B H M E T T R M T O B
A I J F B Q I V V V P W J O P K A O Y T J P I N V K R L G N C V H H Q F M Y E L
O K T Y U B C Y N U N L F G O A P N Q S Z R H P Z D Z V O K B E H J C O T L V U
D J A G W D Z W R Y B L S T L F X V H M W U B S U L B V M P G Q A H L R K X M E
```

Name _____Date_____

Lesson 4   Practice Page 1

Instructions: Fill in the blank with the appropriate word. Refer back to the text portion of the lesson for help.

1. Lipids, like carbohydrates, are formed from three elements: _____,

_____ and _____.

2. There are two main portions to a lipid. The first is a three-carbon component known as -

_____.

3. The second portion consists of chains of carbons and are known as the _____-

_____.

4. There can be up to _____ fatty acid chains per glycerol molecule.

5. A lipid with one fatty acid chain present is known as a _____.

6. A lipid with two fatty acid chains present is known as a _____.

7. A lipid with _____ fatty acid chains present is known as a triglyceride.

8. Fatty acid chains can be classified as being _____ or _____ de-pending upon whether any double bonds are present between the carbon atoms making up the chain.

9. Fatty acid chains which have a double bond present are known as _____ fatty ac-ids.

10. Fatty acid chains which have only single carbon bonds present are known as -

_____ fatty acids.

11. Lipids which are unsaturated usually come from _____ whereas lipids which are saturated come from _____ sources.

12. Unsaturated fats are _____ in form at room temperature while saturated fats tend to be _____ in form at room temperature.

13. Lipids can be converted into fuel (glucose) for living things through the work of -

_____.

14. Because lipids contain so many carbon, hydrogen and oxygen atoms, the relative amount of ener-gy found in one portion of a lipid is about _____ times the amount of energy found in an equal portion of a carbohydrate.

15. Lipids are also found to not mix with water. One can then say that lipids are - _____ in water or _____.

16. Because of their insolubility in water, lipids are useful for living things when water or watery substances need to be contained in one location or another. (True or False)

17. Milk that has all of its fat portion removed is known as _____ milk.

18. Milk that comes straight from the cow has a fat content of _____.

19. The process whereby the fat portion of the milk (cream) is made to remain equally distributed throughout the watery portion of the milk is known as _____.

20. Butter is made from the _____ portion of the milk.

21. Explain how butter can be washed with water.

Below are clues to solve the crossword puzzle on the next page.

Across
2. literally meaning afraid of water
3. form of an unsaturated fat at room temperature
4. when one substance mixes readily with another
6. number of bonds carbon atoms desire to form
10. long chain of carbon atoms making up portion of a lipid
12. when one substance does not mix with another substance
13. form of a saturated fat at room temperature
14. source for most saturated fats
15. process whereby lipid portion of milk is equally distributed throughout watery portion of milk
16. three-carbon portion of a lipid
17. source for most unsaturated fats

Down
1. the quantity of energy a fat has versus the quantity of energy found in an equal portion of carbohydrate
2. literally meaning water loving
4. type of fatty acid when only single carbon bonds are present
5. one of three elements found in a lipid
7. type of fatty acid when a double carbon bond in present
8. one of three elements found in a lipid
9. scientific name for a fat
11. one of three elements found in a lipid

Date

Lesson 4   Practice Page 2

Instructions: On the previous page, you will find clues to solve this crossword puzzle. Refer back to the text portion of the lesson for help.

Name _____Date_____

Lesson 5   Practice Page 1

Instructions: Fill in the blank with the appropriate word. Refer back to the text portion of the lesson for help.

1. Like carbohydrates and lipids, proteins consist of the elements: _____, _____ and _____. However, proteins also always contain the element _____.

2. Proteins consist of smaller components known as _____.

3. There are about _____ amino acids known to man of which there are _____ which are required by humans to live.

4. Amino acids consist of two portions: the _____ portion and then the _____.

5. The common portion is always the same for each amino acid, but the R-group _____ which makes amino acids unique.

6. Of the 20 amino acids necessary for humans to live, many can be made from raw materials in our bodies or other amino acids. However, there are _____ amino acids which cannot be made and therefore must be _____.

7. List those 8 amino essential amino acids:

8. Amino acids are linked to form _____. It is the _____ of these amino acids which form various proteins needed by the body.

9. The process where amino acids are linked together is known as _____ _____ because water is released through the process.

10. A _____ is formed when two amino acids are linking and a - _____ is formed when multiple amino acids are joined together.

11. _____ are responsible for dismantling other proteins to provide a source for components to build the amino acids we need. These _____ also build these new amino acids.

12. The process of breaking down proteins is known as _____ which literally means cutting with _____.

13. Proteins are very large molecules. (True or False.)

14. In general, the primary use for proteins in living things is that they serve as _____ components for the cells of living things.

15. A common protein found in egg whites is _____. This protein is also found in our _____ and is responsible for maintaining appropriate amount of _____ in our bodies.

16. Another important protein found in our blood is _____ which is responsible for _____ formation.

17. And yet another very important group of proteins are the _____ where the prefix immuno– refers to the _____ system and globulin refers to _____.

18. Immunoglobulins work to fight off _____ encountered by the living thing.

19. In mammals, the first milk produced by the mother is full of _____ and has a special name known as _____.

Name _____Date_____

Instructions: Below you will find clues to solve this word find puzzle. Refer back to the text portion of the lesson for help. Note that the words may read forward, backward, up, down or at a diagonal.

1. Proteins are formed from the elements _____, _____. _____ and _____.

2. Proteins are made of smaller units known as _____.

3. Amino acids are made with two main components: the _____ and the _____ which varies.

4. Amino acids get linked together into peptides through the use of enzymes known as _____.

5. The bonds which form the linkages between amino acids are known as _____.

6. Two linked amino acids are known as a _____ whereas many linked amino acids are known as a _____ where poly– means many.

7. The joining of amino acids whereby water is released Is known as _____.

8. It's the _____ of amino acids which determines which protein is produced.

9. The breaking apart of proteins to create a supply of amino acids is known as _____ because water is used to cut the protein.

10. The primary use for proteins in living creatures is their role as structural components. A protein found in blood is _____ aids in blood clotting.

11. Another blood protein is _____ which is the same protein found in egg whites.

12. A final group of proteins are those found in our immune system which are used to fight diseases. This group of proteins are known as the _____.

13. Mammals provide a "megadose" of immunoglobulins through the first milk made for their offspring. This first milk is known as _____.

```
N E N U Y Z E M C P A P X B T V C F J H B D Z P A U O M Y X C T R B R D A G O P
C O O E E S J E E F Y U V C F M A N W K O O E T D F G L M E N J R W E B I G B Q
E J I D G R O P V Y V S X L M M R U L S Q Y I K P A N S C P W P H H Z E R E Z J
P B L T P O T M U R T S O L O C B E O P T F J W B F X X D R T W Y C L L W E J P
U F C I R I R I T G Q E F Z I Q O I J F I W L E C O H A P I N D E N Z Y M E S C
L S T M D O O D G Q Q J N K F Y N K O E Z T B C G E Q G F C R O N P O T P R S L
K Y C A V X P V Y B W F R I X N C W Z M V X D K K P H J A A G Y V O U N I R Y U
D H S N Y D F N F H O U A L R D I S T D B T G B C J L Q T L S G D I Y E J L P U
B E Y G Q E W A O F D W M U E B B K S M A M I N O A C I D S B Y I H X G M N P S
E L E V U N W W I M P U O R G R I N O W D T Z M Z K O P Z O S U X V A O M I M Q
E N R H W X S X J W M W Q P Y M K F N R B Q L J T N M I L I M G M D F R G Z I V
K C S A Z Z F D Y N Q O X U R L X S K B A V X C S I S D E E B J Q I O T X G V O
R C J H U B V Y M B U F C S U S P M R L O M N Y X J N D Z O X I O L N I M C B Z
G A X U B C O P R N P Q B A F Q W I P X C T N C M J E F X Y K C X H K N A P A D
M S Q L C G B D K F H W M N Z J O T Z J S T B A O K H R A B O M Y P I G R O J R
Z E G D O W Z N S V A T S X V J Z X O J H Z L X T Q I X X V T S U A J T C U Y K
F I A O X I B O M L Z N F D L O W X E E S M A D Y U O K U O N Z B A O X M N X N
U N N C W U K B T L V G I Q S K V G S T A S N I L U B O L G O N U M M I B T A H
H I Y P G Q T E I Q W E W Y O O R I X O C S D E Q Z X O S M L C T M F N Q A P R
D Y J K U C G D L Y M B I L E N S O K U Y F F Y K W C S K K B D D F X E U E F V
I A D D U A Y I D U G V W D M T Y I E S H V G Z D E Q Y M C J R H Q Z R T N V V
P J S R M T M T A X O S I H F O L L H F D O E R M S O J E K H E Y M V N H Q X T
E B U C O N H P B W U T D E V W D G I F W Z M X N D A Q M U S T X A M A C L O X
P C O X Y L D E L R P O C P G Q R O Q I P A W Y Y W Y Z B V G A Q Q T W Z L C Y
T S L Q P A Y P F E S N P R J E V N L C O B C P D J F P E C Q W L M D A B P A U
I Y R W N E B S P Q E N O Y I T F K V R J L Q V H Q D G R R U K B L B C P A B C
D B O U I H I Y I U N B J K M U S G R N X P Z O A V T D D I D O B C I P K Y K Y
E W K X D P L K Q S I P E Y E A M X Z T Z K T Y H E O I X I N S F E V N V O G P
Y J Q I H O Z E Y X A N J Z B U P L O Y O N Q B J B N C H E R G H B K D R B O U
U W P H P L S K W J B I Z M L I V S Q I J S B D A V X G B R S Y O R M P F E N E
```

Name _____ Date_____

Lesson 6   Practice Page 1

Instructions: Fill in the blank with the appropriate word. Refer back to the text portion of the lesson for help.

1. All of the uses for carbohydrates, proteins and lipids in living things, whether the building or break-ing down of these molecules, are dependent upon the presence of _____.

2. These enzymes are very dependent upon the _____ of the environment in which they are working.

3. With regard to the symbol used for pH, the lower case p is thought to represent the term _____ while the H stands for _____.

4. Substances which have a desire to get rid of hydrogen ions are said to be _____.

5. Substances which have a desire to accept hydrogen ions are said to be _____.

6. The pH scale tells us the relative strength of acids and bases. As one moves from value to value on the pH scale the relative strength is multiplied _____ times and the scale is referred to as being a _____ scale.

7. An acid with a pH of 3 would be _____ times stronger than an acid with a pH of 5.

8. On the pH scale, acids are less than _____ while bases are greater than _____. Bases taste _____ while acids taste _____.

9. Secretions in our stomachs are considered strong acids. What value on the pH scale would you expect if you measured the pH of these gastric secretions? _____

10. Vinegar or _____ is a _____ acid but still is effective at preserving foods. This is because the _____.

11. Write the pH of the following body substances: Blood _____ Milk _____ Pancreat-ic secretions _____ Cerobrospinal fluid _____ Urine _____

12. By raising or _____ the pH of fluids of living things we can directly affect the en-zymes functioning in that living thing.

13. pH can be measured by colored strips of paper which have a powder affixed to them. These strips of paper are known as _____.

14. Red litmus paper will turn _____ in the presence of a base.

15. Blue litmus paper will turn _____ in the presence of an acid.

Lesson 6   Practice Page 2

Instructions: Below you will find clues
to solve this crossword puzzle.

Across

4. base used to leaven cookies

6. greater than seven on the pH scale

8. acids taste

9. bases taste

11. the pH scale's difference between
values

13. pH of blood

14. color of litmus paper in presence of
acid

Down

1. scientific name for vinegar

2. what the H stands for in pH

3. pH of urine

4. color of litmus paper in presence of
base

5. pH strips capable of testing range of
pH values

7. less than seven on pH scale

10. enzymes are dependent upon _____
to function correctly

12. what the p stands for in pH

Name _____Date_____

Lesson 7   Practice Page 1

Instructions: Fill in the blank with the appropriate word. Refer back to the text portion of the lesson for help.

1. The study of cells is known as _____.

2. The scientist who first made the discovery of cells was _____ in _____. He gave cells their name because it is said he though they resembled _____.

3. The "little organs" within cells are known as _____.

4. Cells are covered with a "skin" known as the _____ or _____.

5. The cell membrane consists of a _____ bilayer.

6. The phosphate groups are _____ while the lipid layer is _____.

7. This allows the membrane to be able to control the passage of _____ .

8. Cell membranes are said to _____ meaning that the cell has the ability to allow certain substances to pass but not others.

9. Plants have an extra layer of protection around their cells. This layer is known as the _____.

10. Cell walls are made of _____ which are long strings of _____ molecules.

11. Cells walls function like _____ for plant cells giving them strength and support.

12. Trees which have thin cell walls and tend to grow quickly are known as _____ while trees which have thick cell walls and tend to grow slowly are known as _____.

13. The cell organelle which functions like the brain of the cell is the _____.

14. The nucleus, depending upon which creature it is in, may or may not have a covering membrane. This membrane is known as the _____ or _____.

15. Organisms which DO have a nuclear membrane are identified as being _____ .

16. Organisms which do NOT have a nuclear membrane are identified as being

_____.

17. An example of an organism which is a prokaryote is a _____.

18. The structures within the nucleus of a cell which hold the instructions for the cell are the

_____.

19. Organisms are specific for the _____ of chromosomes they have and for the infor-mation found on each chromosome.

20. Humans have _____ pairs of chromosomes for a total of _____ chromosomes in most of their cells.

21. Within chromosomes we find subunits or "recipes" which are known as _____.

22. Genes code for _____ in the organism.

23. The organelle of cells which functions like a power plant is the _____.

24. The fuel utilized by the power plants of cells is _____.

25. The chemical formula for glucose is _____.

26. Mitochondria are shaped like _____.

27. The inner membrane of a mitochondrion is known as _____ and this is where _____ takes place.

28. Respiration is the process whereby glucose is converted into _____ which is in the form of _____s.

29. ATP is the abbreviation for _____.

30. In the presence of oxygen, one glucose molecule can yield _____ ATPs.

31. This form of respiration is known as _____ respiration.

32. Where there is little or no oxygen present, one glucose molecule can only yield _____ ATPs.

33. Respiration with little to no oxygen present is known as _____ respiration.

34. The two main uses of ATPs are to allow for _____ and _____ produc-tion.

Name _____Date_____

Lesson 7   Practice Page 2

Instructions: On the next page, you will find clues to solve this crossword puzzle. Refer back to the text portion of the lesson for help.

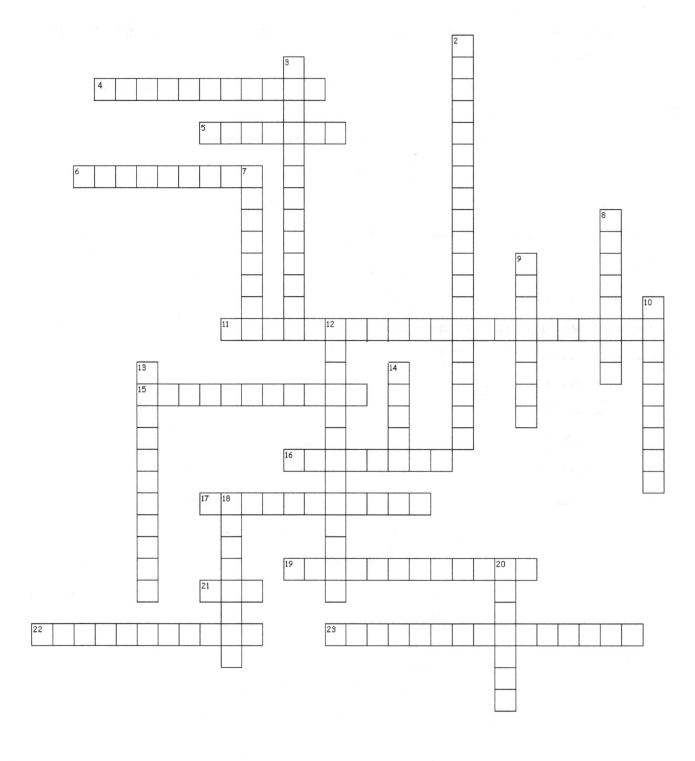

Clues for Crossword Puzzle

Across

4. loves water

5. fuel used by mitochondrion of cell

6. what discover of cells thought they looked like

11. what ATP stands for

15. hates water

16. total number of chromosomes in most of a human's cells

17. aerobic respiration makes this many ATPs.

19. "skin" of cells

21. anaerobic respiration results in this many ATPs.

22. process whereby glucose is converted into ATPs.

23. chemical formula for glucose

Down

2. scientific term which describes how cell membranes are composed of two layers of phosphates and fat compounds

3. "power plant" of cell

7. thin cell wall of pine or fir

8. skeleton of plant cell

9. study of cells

10. carbohydrate which makes cell wall

12. means: allows some things to pass but not others

13. colored bodies which are like chapters in the cell's cookbook

14. "recipes" found in the chromosomes

18. thick cell wall of oak or maple

20. "brain" of cell

Name _____Date_____

Lesson 8   Practice Page 1

Instructions: Fill in the blank with the appropriate word. Refer back to the text portion of the lesson for help.

1. The organelle of cells which is responsible for packaging of products for export from the cell is the _____.

2. When a portion of golgi body is filled with product and ready to be shipped from the cell, a portion known as a _____ breaks free to moves toward the periphery of the cell.

3. The organelle of a cell which functions primarily as storage or, in some cases, locomotion, is the _____.

4. The part of a cell which functions like a transportation system is the _____ or _____ for short.

5. There are two types of ER: _____ and _____ ER.

6. The roughness is rough ER is due to the _____ present.

7. The function of ribosomes are to build _____ from available _____.

8. Another role performed by ER is the folding of _____.

9. There are four degrees of protein folding: _____, _____, _____ and _____.

10. The functionality of a protein is dependent upon the way it is _____.

11. Sarcoplasmic reticulum is found in _____ cells and is important in the regulation of _____.

12. _____ are like a microscopic skeletal system for cells. They can enable movement, too.

13. Microtubules which make up centrioles are important in _____.

14. _____ are organelles found in green plants which are capable of the process of _____.

15. The raw materials for photosynthesis include: _____, _____, and energy from the _____.

16. The product of photosynthesis is _____ which has the chemical formula of _____.

17. This glucose can be used by _____ or can be "stolen" by another living creature to be used as a _____.

18. A waste product from cellular _____ is _____ which becomes a resource for plant use in photosynthesis.

19. Plants and creatures which undergo cellular respiration are _____ upon each other.

Clues for crossword puzzle on next page.

Across

6. fourth degree of protein folding

9. transportation system of cell

12. second degree of protein folding

13. ER with ribosomes

17. formula for glucose

20. little protein building factories found on rough ER

Down

1. photosynthesis factory of plants

2. reticulum of muscles

3. closet for cell

4. process of making glucose from carbon dioxide, water and sunlight

5. this element is controlled by sarcoplasmic reticulum

7. third degree of protein folding

8. we are _____ upon plants and they are _____ upon creatures like us

10. works like a UPS delivery person

11. tiny bones of cells

14. first degree of protein folding

15. byproduct of photosynthesis but useful for animals

16. important in cell division

18. ER without ribosomes

19. package ready for shipment from cell

Name _____Date_____

Lesson 8   Practice Page 2

Instructions: Find clues to solve this puzzle on the previous page.

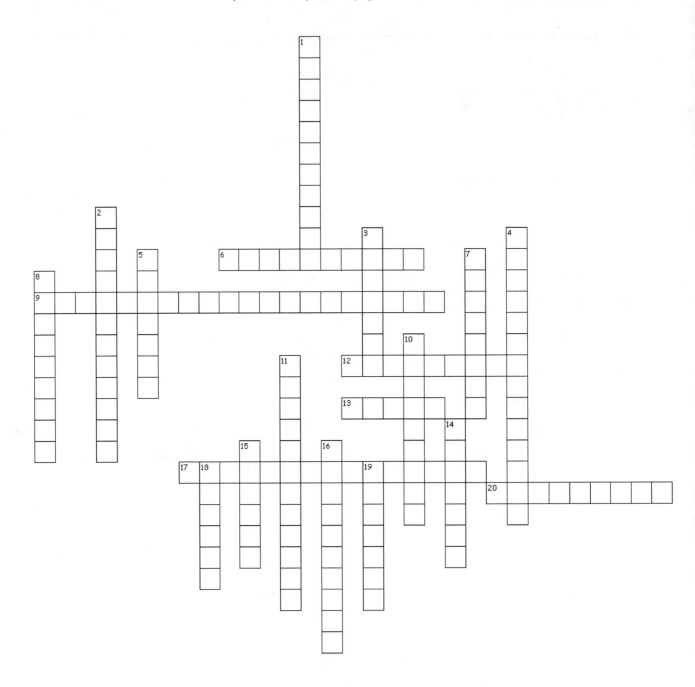

Name _____Date_____

Lesson 9   Practice Page 1

Instructions: Fill in the blank with the appropriate word. Refer back to the text portion of the lesson for help.

1. Living things which are made up of many cells grow in size by _____ the number of cells from which they are made.

2. The process of cell division is known as _____.

3. The phase of a cell's life cycle when it is going about its assigned job is known as
_____.

4. In animals and humans stimulation to divide usually comes in the form of a
_____.

5. Hormones are substances usually made by _____ in one part of the body and have their effects in other parts of the _____ being carried there by the
_____.

6. Growth hormone is made by the _____ which is located
_____.

7. The cell that receives the signal to divide is referred to as the _____ cell and the two cells which result from the division are known as the _____ cells.

8. The first stage of mitosis is known as _____.

9. In prophase the _____ which was originally uncoiled while being used by the cell, _____ or bunches up and becomes visible again as chromosomes.

10. The _____ also begins to break apart during this stage.

11. The second main event of prophase is that the chromosomes of the parent cell must
_____.

12. When copying of the chromosomes does not happen correctly, a _____ is said to have occurred.

13. The second stage of mitosis is known as _____. During this phase the chromosomes align themselves along a _____ within the cytoplasm of the cell. _____ fibers attach themselves to the chromosomes.

14. The next stage of mitosis is known as _____.

15. During anaphase, the spindle fibers _____ pulling their attached

_____ to opposite poles of the cell.

16. The final stage of mitosis is known as _____.

17. During telophase the chromosomes begin to _____ and the cell membrane begins to _____ eventually allowing for the cell to _____ into _____ cells.

18. The process whereby the cell membrane presses inward to create two daughter cells is known as _____.

19. The two "new" daughter cells return back to _____ where they continue their jobs.

Name _____ Date_____

Instructions: Below you will find clues to solve this word find puzzle. Refer back to the text portion of the lesson for help. Note that the words may read forward, backward, up, down or at a diagonal.

1. Phase of mitosis where chromosomes are moving to opposite poles of cell.

2. Part of chromosome where spindle fiber attaches.

3. When chromosome is uncoiled and being used by the cell, this genetic material is called

_____.

4. During prophase, the uncoiled chromatin will _____ to form chromosomes.

5. _____ of the chromosomes are made during prophase.

6. During telophase, the cell membranes pinches inward, a process known as _____.

7. Parent cells divide into two _____.

8. Phase of cell life cycle where "regular job" is taking place.

9. Stage of mitosis where chromosomes line up along central plate or equator of cell.

10. Cell division = _____

11. _____ divide to form daughter cells.

12. During metaphase, the pairs of chromosomes line up along a central _____ or equator of the cell.

13. During anaphase the chromosomes get pulled to opposite _____ of the cell.

14. During _____, chromatin coils up to form chromosomes and the chromosomes duplicate themselves.

15. _____ attach to chromosomes during metaphase to eventually pull the chromosomes to opposite poles of cell.

```
P N F S S B S L P P V A T O E
R E H P P L G I U O N N D H S
O R C I M V L L S A L A Y E A
P G X N C E I E P O U E R E H
H S N D P O T H C G T E S W P
A E P L C J A A H T M I A X R
S I X E L S W T P O N K M O E
E P I F E X E K R H L E E J T
M O A I G R T T N S A H R S N
W C J B C L N F V E X S C A I
H T V E T E E S A H P L E T P
L J L R C C H R O M A T I N
Y L C S Q S G C G P L A T E H
S H T E X N A V Z G H W P S D
S B S I S E N I K O T Y C R T
```

Name _____Date_____

Instructions: Fill in the blank with the appropriate word. Refer back to the text portion of the lesson for help.

1. Chromosomes must _____ themselves in order for each daughter cell to have its own set of chromosomes.

2. Duplication of chromosomes is made possible through the function of _____ which work to either _____ or "matchmake" the appropriate parts of the DNA.

3. DNA is the abbreviation for _____.

4. The carbohydrate component of DNA is a _____ sugar.

5. The nucleotide portion consists of the _____ sugar, a _____ group and then a pair of _____.

6. The "rung" portion of the DNA ladder is made up of pairs of _____.

7. There are two groups of bases: the _____ and the _____.

8. The two purines are _____ and _____.

9. The two pyrimidines are _____ and _____.

10. _____ always bonds with _____ while _____ always bonds with _____.

11. The bonds between the bases are weak bonds known as _____ bonds.

12. Enzymes _____ the hydrogen bonds exposing the base pairs.

13. Matchmaking enzymes bring in _____ in the area joining them according to the matching rules.

14. Eventually, _____ new _____ strands of DNA are formed; one strand for each _____ cell.

Lesson 10   Practice Page 2

Instructions: On the next page, you will find clues to solve this crossword puzzle. Refer back to the text portion of the lesson for help.

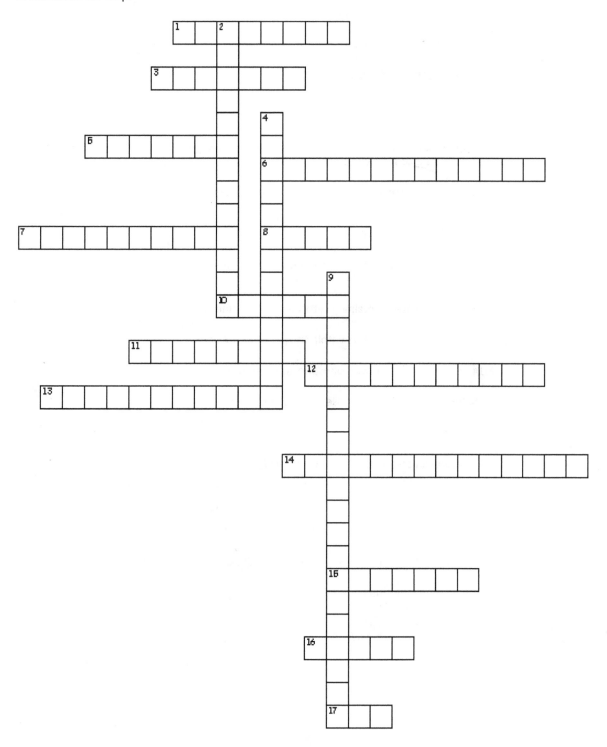

Across

1. always bonds with guanine

3. group of bases containing adenine and guanine

5. always bonds with thymine

6. "new" cells which each need their own copy of the DNA cookbook.

7. Chromosomes must be _____ so that each daughter cell gets a complete copy of all DNA.

8. These are the recipes within each chapter of the genetic cookbook.

10. The sugar found in DNA.

11. The stronger bond between the ribose sugar and phosphate group of a nucleotide is a _____ bond.

12. group of bases containing cytosine and thymine

13. These are the chapters within the genetic cookbook.

14. Part of nucleotide with phosphorus present.

15. these unzip DNA to expose the base pairs

16. Parts of nucleotide.

17. Abbreviation for deoxyribonucleic acid.

Down

2. Shape of DNA discovered by Watson and Crick.

4. Type of bond found between base pairs that is unzipped by enzymes.

9. Chromosomes are made of long strands of _____.

Name _____Date_____

Lesson 11   Practice Page 1

Instructions: Fill in the blank with the appropriate word. Refer back to the text portion of the lesson for help.

1. The making of proteins is known as protein _____.

2. The "recipe" for making a protein correctly is found in one's _____.

3. There are four nitrogen bases found in DNA and they are _____, _____, _____ and _____.

4. And, it is the _____ of these bases which determines the type of protein that can be synthesized.

5. Before a protein can be made, the particular section of DNA must be _____ by _____ to expose the base pairs.

6. After the base pairs are exposed, _____ can be formed to make a "copy" of the sequence of the bases.

7. Unlike DNA, RNA is _____ stranded and contains the base _____ instead of _____.

8. The RNA which forms on the exposed DNA bases takes the name _____ or mRNA for short.

9. The formation of mRNA takes place in the _____ of the cell whereas protein synthesis takes place out on/in the _____ of the cell.

10. More specifically, proteins get built in the _____ found on the _____ endoplasmic reticulum

11. Out at the ribosome, available amino acids are "tagged" with _____ or tRNA for short.

12. The tRNA is specific for each _____ and will therefore only fit the appropriate sequence of the mRNA that has arrived at the ribosome. The tRNA works like a _____ that fits the mRNA "lock."

13. The tRNA is in sets of _____ base pairs and is called an _____.

14. Specific codons code for specific _____ acids.

15. As the amino acids get aligned to the mRNA, _____ work to link together forming _____ bonds. Long strands of peptides form in this way.

16. The process whereby mRNA is made from a section of DNA is called _____ where the process where tRNA assembles amino acids in the ribo-some is known as _____.

17. When one's DNA is altered or damaged, the proteins made from this damaged DNA may not _____ correctly. A change in the DNA is called a _____ mutation.

18. Things that cause mutations are called _____ or _____.

19. An example of a mutagen is _____ which is energy that can come from the sun or an elemental source.

20. Cells which have had their DNA altered may become _____ cells and change their behavior.

21. Cancer treatments may include bombarding cancer cells with radiation in an attempt to cause _____ in the DNA of those cells leading to their destruction.

Name _____ Date_____

Lesson 11   Practice Page 2

Instructions: Below are 17 words or phrases that we've introduced in this lesson. Their spellings have been scrambled. Rearrange the letters to spell these words and place them into the boxes. Transfer the letters in the numbered boxes to the "finale" at the bottom. If you need hints on unscrambling the words, turn the page.

QESNECUE ☐☐☐☐☐☐☐☐ (16)

MGNERSERNASE ☐☐☐☐☐☐☐☐☐☐☐☐ (18, 24, 25)

SENFARANTRR ☐☐☐☐☐☐☐☐☐☐☐ (5, 21)

NAD ☐☐☐ (19)

BOIRESMOS ☐☐☐☐☐☐☐☐☐ (23, 3, 20)

LUUSECN ☐☐☐☐☐☐☐ (8)

MIENIRASOLLUCTEDUCPM ☐☐☐☐☐☐☐☐☐☐☐☐☐☐☐☐☐☐☐☐ (1, 17)

GEURHOR ☐☐☐☐☐☐☐ (2, 12)

ZEYNMES ☐☐☐☐☐☐☐ (22, 9)

DOPSBDETPENI ☐☐☐☐☐☐☐☐☐☐☐☐ (4)

MANODSICIA ☐☐☐☐☐☐☐☐☐☐ (6)

SONCDO ☐☐☐☐☐☐ (10)

RANSOITLNAT ☐☐☐☐☐☐☐☐☐☐☐ (7)

SRNAITOINTRPC ☐☐☐☐☐☐☐☐☐☐☐☐☐ (11)

TOTMANSIU ☐☐☐☐☐☐☐☐☐ (14)

ANRIATIOD ☐☐☐☐☐☐☐☐☐ (15)

REACNC ☐☐☐☐☐☐ (13)

☐☐☐☐☐☐☐ ☐☐☐☐☐☐☐☐☐ ☐☐ ☐☐☐☐☐☐☐ .
1 2 3 4 5 6 7   8 9 10 11 12 13 14 15 16   17 18   19 20 21 22 23 24 25

39

Hints for solving the scrambled words puzzle are found below. Note that these hints are not in the same order as the puzzle words.

Transcribing molecules

These get linked together to form proteins.

It is the _____ of bases which determines the protein synthesized.

These make smooth ER not smooth.

Sets of threes.

Location of DNA

ER

ER that is not smooth.

Tags or keys found on amino acids

Scissors or matchmakers

Bonds between amino acids.

When genes get messed up.

Rays from the sun.

Disease where cells are out of control.

Ladder-like molecule with all information.

mRNA does this job

tRNA does this job

Name _____Date_____

Lesson 12   Practice Page 1

Instructions: Fill in the blank with the appropriate word. Refer back to the text portion of the lesson for help.

1. The process by which living organisms create more complete living organisms is known as

_____.

2. Organisms with two parents have pairs of _____, one member of each pair coming from each _____.

3. Humans have _____ pairs of chromosomes for a total of _____.

4. The variation in a trait is known as an _____. Examples of alleles for eye color could be: _____.

5. When it comes to alleles, chocolate is to ice cream as _____ is to the gene for hair color.

6. Living creatures which have pairs of chromosomes are said to be _____ or 2N.

7. Almost all cells in humans and animals are 2N except for the _____ cells or

_____.

8. These cells (sex cells or gametes) are _____N or haploid.

9. The cells which eventually become gametes are known as the _____

_____ and originally are _____N.

10. The primordial sex cells undergo the process of _____ to reduce the number of chromosomes from 2N to _____.

11. In human males, one primordial sex cell will result in _____ sperm cells.

12. In human females, one primordial sex cell will result in _____ ovum and three cells which do _____ survive.

13. Human males continue to create more primordial sex cells throughout their lives, while human females are born with the total number of primordial sex cells they will have their entire lives. (True or False)

14. In males, the process of creating sperm cells takes place in the _____.

15. In females the process of creating ova takes place in the _____.

16. The joining of the 1N sperm cell with the 1N ovum to create a 2N individual is known as

_____.

17. Reproduction which involves two parents, a male and female each contributing a single cell for the new offspring, is known as _____ reproduction.

18. Reproduction which involves only one parent organism is known as _____ reproduction.

19. In _____ reproduction there is a great possibility for an assortment or mixing of genetic information from each parent.

20. In _____ reproduction, the offspring carries the exact same genetic information as the parent. There is not any change.

21. Bacteria utilize a means of asexual reproduction whereby the "parent" bacteria creates a copy of its genetic material and then promptly splits into two equal halves, each with its own copy of genes. This method of reproduction is known as _____.

22. Another means of asexual reproduction is where genetic material is isolated near the outside of the organism. The cell membrane encircles the genetic material and other cellular organelles forming a bud. The bud eventually breaks free. This method is known as

_____.

23. Some plants, like strawberries and spider or airplane plants create "baby" plants by sending out special stems called _____ which eventually touch ground forming roots and new leaves. This method that plants use is called _____

_____.

24. The plants which result from vegetative propagation will have genetic information unlike the parent plant. (True or False)

25. Ferns and mushrooms reproduce by producing spores. This method of reproduction is known as

_____.

26. Some creatures, like starfish for example, are able to regenerate whole new living organisms from broken parts. This form of reproduction is known as _____.

27. Two parent involvement = _____ reproduction.

28. One parent involvement = _____ reproduction.

29. Sexual reproduction = mixing of genetic information in offspring. (True or False)

30. Asexual reproduction = no mixing of genetic information in offspring. (True or False)

Name _____Date_____

Lesson 12   Practice Page 2

Instructions: On the next page, you will find clues to solve this crossword puzzle. Refer back to the text portion of the lesson for help.

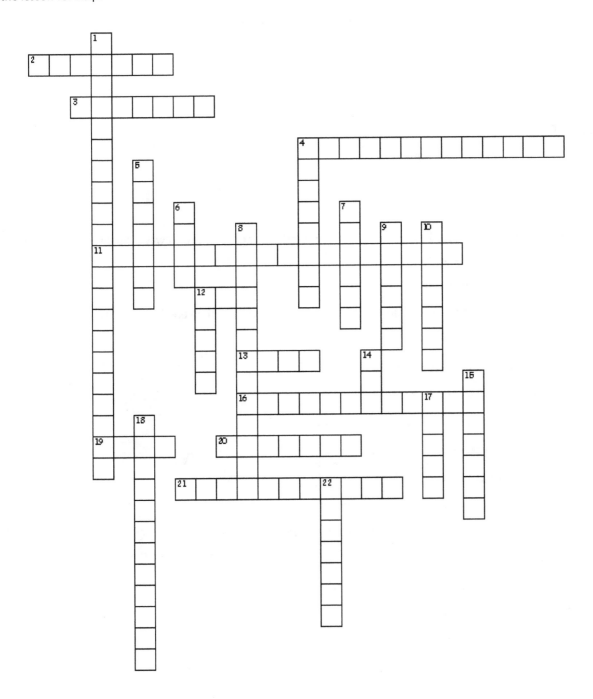

Across

2. term used for sex cells

3. process where a diploid cell becomes haploid

4. joining of ovum and sperm to create new individual

11. cells which are 2N and destined to undergo meiosis to become 1N 12. female gametes

13. number of gametes created in males for each primordial sex cell

16. process of reproducing through the formation of spores

19. haploid

20. having pairs of chromosomes

21. number of pairs of chromosomes in humans

Down

1. means of reproducing often employed by plants where specialized stems or roots grow away from parent plant with small "baby" plant forming on end of stem

4. total number of chromosomes in humans

5. asexual form of reproduction where a small portion of parent cell emerges and then breaks free from parent cell

6. diploid

7. reproduction where two parents are involved

8. form of reproduction where parent cell divides into two new cells

9. location in males where gametes are produced

10. variations of a trait on a chromosome

12. location in females where gametes are produced

14. number of gametes created in females for each primordial sex cell

15. reproduction where only one parent creates new living organism

17. male gametes

18. process where a living organism creates more living organisms like itself

22. having half the number of chromosomes

Name _____ Date_____

Lesson 13   Practice Page 1

Instructions: Fill in the blank with the appropriate word. Refer back to the text portion of the lesson for help.

1. The biologist who discovered many concepts of genetics was named _____.

2. One of Mendel's biggest contributions was that traits are _____

_____.

3. The record of genes found in a living organism is referred to as the _____.

4. The results of the genotype that can be observed whether physical attributes or behaviors is referred to as the _____.

5. The possible variations of a gene are referred to as _____.

6. We can think of alleles as the possible _____ available for ice cream or cookies.

7. The Law of _____ says that, in meiosis, one chromosome from each parent primordial sex cell will go to each resulting gamete.

8. The Law of _____ says that traits get distributed independently from other.

9. When looking at the genotype of the offspring that was created through sexual reproduction, we find _____ present, one from the male parent and one from the female parent.

10. While both alleles are present, only one gets expressed. The _____ allele, if present, is one that always gets expressed.

11. The allele that is present, but does not get expressed is the _____ allele.

12. When describing genes, one usually uses _____. To indicate dominant alleles, one uses _____ case letters and recessive alleles are indicated by _____ case letters.

13. When dominant alleles are inherited from both parents, for example HH, the individual is said to be _____ dominant for the H allele.

14. When recessive alleles are inherited from both parents, for example hh, the individual is said to be _____ _____ for the H allele.

15. When an individual inherits a combination of a dominant and recessive allele for a specific trait,

that individual is said to be _____ for that trait.

16. Suppose hair color is indicated using the letter "H." Suppose, also, that black hair color is dominant and red hair color is recessive. If this is the case, what color hair would these individuals have:
HH _____ Hh_____ hh_____

17. If Tom has black hair (from question 16 above,) what are his possible genotypes?
_____ or _____

18. If Tom has red hair, what is his only possible genotype? _____

19. If Tom is heterozygous for black hair color, what must his genotype be? _____

20. If Tom is homozygous dominant for black hair color, what must his genotype be? _____

21. If all of Tom's biological children have black hair, regardless of the hair color of his wife, what must Tom's genotype be? _____

22. If Tom has some children with black hair and some with red hair, what must his genotype be?
_____

23. If Tom has the genotype Hh and his wife has the genotype Hh, do they have the chance of having a child with red hair? _____

24. If Tom has red hair and he marries his wife who has black hair, could they ever have children with red hair? _____ How can this be? _____

25. If Tom has red hair and he marries his wife who also has red hair, will they ever have children with black hair (according to our "story" of hair color dominance)? _____

26. If Tom has black hair and his wife has black hair, could they ever have the chance of having a child with red hair? _____ If so, how could this occur? _____

_____.

27. Suppose Tom is homozygous dominant for black hair color and his wife is homozygous dominant for black hair color. Will they ever have the chance of having children with red hair?
_____ Why or why not? _____.

28. The situation where the phenotype of an individual appears to be the result of sharing of two dominant alleles is known as _____.

29. When one finds that certain traits, such as down color of chicks, is found only in one gender, the trait is said to be _____.

Name _____Date_____

Lesson 13   Practice Page 2

    Suppose you have some show-quality cats that you would like to breed to be able to sell their kittens. While all of the kittens bring good profits, the ones with long hair bring the most returns. Suppose that hair length in cats utilizes the letters "L" and "l" for the genotype. Suppose, also, that long hair length is a recessive trait. Look at the following situations and predict the possible outcomes for hair length in your kittens. Use the Punnet squares to support your predictions. Male cats are referred to as toms and females are referred to as queens.

A. Your tom is homozygous recessive (l l) and your queen is heterozygous (Ll) for hair length.

Number of kittens with

Long hair: _____

Short hair: _____ out
of every four kittens.

B. Your tom is heterozygous (Ll) and your queen is homozygous (l l) recessive for hair length.

Number of kittens with

Long hair: _____

Short hair: _____ out
of every four kittens.

C. Your tom is heterozygous and your queen is heterozygous for hair length.

Number of kittens with

Long hair: _____

Short hair: _____ out of every four kittens.

D. Your tom is homozygous recessive and your queen is homozygous recessive.

Number of kittens with

Long hair: _____

Short hair: _____ out of every four kittens.

E. Your tom is homozygous dominant and your queen is homozygous dominant.

Number of kittens with

Long hair: _____

Short hair: _____ out of every four kittens.

Which situation (question letter, A-E) would potentially be the best money-making opportunity for you? _____

Name _____Date_____

Lesson 14   Practice Page 1

Instructions: Fill in the blank with the appropriate word. Refer back to the text portion of the lesson for help.

1. The study of the form of living things is known as _____.

2. The study of grouping organisms based upon similarities and differences is known as

_____.

3. While many scientists have contributed to the study of taxonomy, one in particular can be credited for a tremendous amount of early work. His name was _____.

4. The most important reason living things are placed into groups and given specific names, is that scientists need to _____ accurately about what they learn about living things with each other.

5. Living things all fall into one of five major groups known as kingdoms. These five kingdoms are:

_____, _____, _____,

_____ and _____.

6. Examples of the Animalia kingdom might include: _____, _____ and

_____.

7. Examples of the Plantae kingdom might include: _____, _____ and

_____.

8. Examples of the Fungi kingdom might include: _____, _____ and

_____.

9. Bacteria and viruses are members of the _____ kingdom.

10. Unicellular, swimming creatures are members of the _____ kingdom.

11. The classification division immediately beneath the kingdom level is the _____ level.

12. Humans fall into the phylum _____ because of the presence of a

_____.

13. Why do dogs also belong to this phylum? _____

14. Beginning with kingdom and phylum, list the remaining levels of classifications for living things: kingdom, phyum, _____, _____, _____,

_____ and _____.

15. The level of classification which serves as the "first name" of an organism is _____ while the

_____ level serves as the "last name."

16. There are rules for writing the scientific name of organisms. The _____ portion is capitalized while the _____ portion is not. If hand-written, they are to be _____ and if typed, _____.

17. In this lesson, we found that size was a determining factor used in grouping in members of the cat family. In horses and cattle, a determining feature was the number of _____ on each foot. Cattle are classified as being _____ while horses are classified as being _____. 18. How would a whitetail deer be classified? _____. How about a gray zebra? _____

19. Using the Catalogue of Life resource, find the scientific names for the following living organisms:

    A. Eastern Cottontail rabbit: _____

    B. White oak: _____

    C. Oyster mushroom: _____

    D. Domestic dog: _____

    E. House fly: _____

    F. Red squirrel: _____

20. Using the Catalogue of Life resource, find the common name for the following living organisms:

    A. *Bos taurus* _____

    B. *Sus scrofa* _____

    C. *Canis lupus* _____

    D. *Julglan nigra* _____

    E. *Homo sapien* _____

Name _____Date_____

Lesson 14   Practice Page 2

Instructions: Below, you will find clues to solve this crossword puzzle. Refer back to the text portion of the lesson for help.

Across

3. Kingdom that we find puffballs.

4. Fellow who did much of early taxonomy.

6. Kingdom of tiny swimmers.

8. Chordata is an example

11. All members of the phylum Chordata have _____

13. Lions and tigers and bears, oh my!

15. study of groupings of living things based upon similarities and differences

Down

1. "highest" category of taxonomic hierarchy

2. Sunflowers belong to this kingdom.

5. Species name is written using _____

7. Scientific name format when type-written.

9. Genus name is always written using _____

10. Kingdom that we find bacteria.

12. Phyla are divided into _____.

14. study of forms and shapes of living things

Name _____Date_____

Match the word or phrase in Column A with the <u>best</u> word or phrase in Column B. Write the letter from Column B beside the number in Column A. The first question has been completed for you!

Column A	Column B
__A__1. Kingdom Animalia	A Includes all animals
_____2. Rotifera	B Phylum Platyhelminthes
_____3. Phylum Mollusca	C Phylum Porifera
_____4. Class Cestoda	D Phylum Cnidaria
_____5. Characterized by having namatocysts	E Class Scyphozoa
_____6. Sea anemones belong in this class	F Class Anthozoa
_____7. Flukes are a member of this class.	G Class Trematoda
_____8. Flat worms	H Have a little crown of twirling cilia.
_____9. Comb jellyfish belong to this phylum.	I Phylum Ctenophora
_____10. True jellyfish	J Tapeworms
_____11. Phylum Annelida	K Spiny-headed worms
_____12. Phylum Bryozoa	L Moss animals
_____13. Phylum Brachiopoda	M Upper and lower shell
_____14. Sponges	N Have a mantle which creates a shell
_____15. Class Bivalvia	O Includes snails, slugs and whelks.
_____16. Class Gastropoda	P Clams, oysters and scallops
_____17. Class Diplopoda	Q Members include the squid and octopus
_____18. Chitons	R Shell made of eight valves known as plates
_____19. Phylum Acanthocephala	S Segmented worms like earthworms
_____20. Class Insecta	T Segmented body parts and jointed legs.
_____21. Class Arachnida.	U Spiders
_____22. Subphylum Crustacea	V Two pairs of antennae and ten legs.
_____23. Class Chilopoda	W Centipedes
_____24. Class Cephalopoda	X Insects
_____25. Phylum Arthropoda .	Y Millipedes
_____26. Phylum Echinodermata.	Z Pentaradially symmetrical.
_____27. Phylum Hemichordata.	AA Acorn worms
_____28. Phylum Chordata	BB Animals with backbones.

Name _____ Date _____

Lesson 16   Practice Page 1

Match the word or phrase in Column A with the <u>best</u> word or phrase in Column B. Write the letter from Column B beside the number in Column A. The first question has been completed for you!

Column A	Column B
__A__ 1. Phylum chordate	A Phylum of animals possessing a notochord, gill slits and a postanal tail (at one point in development).
_____ 2. Subphylum Cephalochordata	B Lancelets
_____ 3. Class Chondrichthyes	C No-jawed fish; lamprey and hagfish
_____ 4. Order Trachystoma	D Sharks, rays and skates
_____ 5. Class Osteichthyes	E Fishes that have bony skeletons
_____ 6. Order Anura	F Aquatic at some point in their lives and then become terrestrial
_____ 7. Class Amphibia	G Amphibians with no legs, blind worms
_____ 8. Order Apoda	H Newts and salamanders
_____ 9. Class Agnatha	I Amphibians with no tail, frogs, toads.
_____ 10. Class Aves	J Rough-mouthed amphibians
_____ 11. Class Reptilia	K Covered with scales, only have lungs, internal fertilization
_____ 12. Order Rhynchocephalia	L Tuatara of New Zealand
_____ 13. Order Edentata	M Order include turtles and tortoises.
_____ 14. Order Squamata	N Includes crocodiles and alligators
_____ 15. Order Chiroptera	O Snakes and lizards
_____ 16. Order Crocodilia	P Class includes all birds
_____ 17. Order Urodela	Q Warm-blooded, milk-producing, hairy animals
_____ 18. Order Insectivora	R Included the egg-laying mammals: platypus and echidna
_____ 19. Class Mammalia	S Pouch-possessing mammals
_____ 20. Order Chelonia	T Moles and shrews
_____ 21. Order Rodentia	U flying mammals, bats
_____ 22. Order Monotremata	

_____25. Order Cetacea

_____26. Order Carnivora

_____27. Order Sirenia.

_____28. Order Artiodactyla

_____29. Order Proboscidae.

_____30. Order Lagomorpha

_____31. Order Pinnipedia

_____32. Order Perissodactyla

_____33. Order Primates

V Peg-like teeth armadillo and sloth

W Includes whales, porpoises and dolphins.

X Members of this order include the manatees

Y Elephants

Z Meat-eaters; foxes, wolves, dogs, cats

AA Seals, walruses

BB Horses, donkeys, rhinos

CC Include the cattle, sheep, deer, giraffe

DD Monkeys, lemurs, gibbons, orangutans, chimpanzee, gorillas and humans.

EE Includes rabbits, hares and the pikas

FF Rats, mice, squirrels

GG The pangolins

Name _____Date_____

Lesson 17   Practice Page 1

Please fill in the blank(s) for each statement below. Use your lesson to help you.

1. Plants, through the process of _____ are able to take carbon dioxide, water and sunlight to produce _____ which is fuel for the plant.

2. Plants are considered to be _____ in that they are able to produce their own food supply. Living things which depend upon other living things as food are said to be

_____.

3. Photosynthesis in plants takes place in the _____ of the cells of plants.

4. Unlike the cells of animals, plants have _____ which are made up primarily of _____ which are long chains of glucose molecules.

5. Taxonomists divide plants into two large groups based upon the presence of _____ tissue.

6. Vascular tissue which carries water is known as the _____ while vascular tissue which carries food (glucose) around the plant is known as _____.

7. Instead of being divided into phyla, plants within the Kingdom Plantae are divided into

_____.

8. There is one division of plants in the non-vascular division. This is the Division

_____.

9. Three classes are present within this division. The plants found in these classes include the _____, _____ and _____.

10. The Division Psilophyta has _____-like leaves. The most common member is the

_____.

11. The Division Sphenophyta includes plants known as _____ which have hollow stems and contain _____.

12. The Division Pterophyta includes _____ which have well developed underground rhi-zomes.

13. Cycads belong to the Division _____ and have seeds not enclosed in

_____.

14. The lone member of the Division Gingkophyta is the _____.

15. The _____ plant is a member of the Division Gneophyta has two very long leaves and a very short trunk.

16. Trees such as pines, firs and spruces belong to the Division _____. They are commonly referred to as _____ and have their seeds in _____.

17. Their leaves are like _____.

18. Plants which have flowers and seeds enclosed in a fruit belong to the Division

_____.

19. There are two classes found within the Division Anthophyta. The first includes plants with one cotyledon and are commonly referred to as _____. Examples of these include: _____

20. Flowering plants with seeds having two cotyledons are commonly referred to as _____ and examples are: _____.

21. Flowers with both male and female structures present are known as _____ flowers whereas flowers with only one sex of reproductive structure are known as _____ flow-ers.

22. The male parts of a flower are the stem-like _____ with the pollen-producing part known as the _____.

23. The female part of a flower includes the _____ which has a stem-like portion known as the _____ and the sticky top known as the _____.

24. Pollen carry two types of cells: one which forms the _____ while the second divides to create two _____.

25. The pollen tube allows the _____ cells to join the _____ present in the ovary of the flower. From this joining process, many cell divisions result and the seed, enclosing the plant embryo, develops.

Name _____Date_____

Lesson 17   Practice Page 2

Use the clues on the next page to solve this crossword puzzle regarding the plant kingdom.

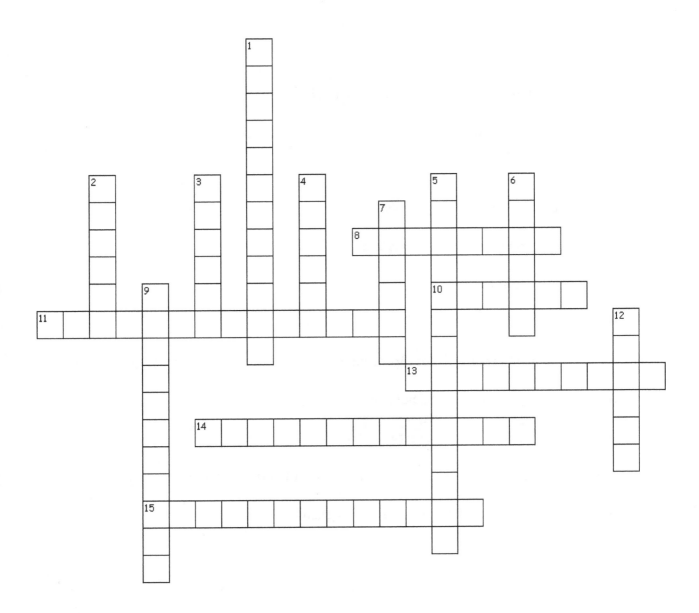

Clues:

Across

8. plants having parallel leaf veins and seeds with one cotyledon

10. structures within ovary which are female contribution to new plant embryos

11. tissue of plants which is capable of transporting water and food products

13. division of plants which includes flowering plants

14. division of plants which includes pines, spruces and firs

15. living things which depend upon other living things as a food supply

Down

1. organelles which carry out photosynthesis

2. plants with branched leaf veins and seeds with two cotyledons

3. sticky end of female pistil where pollen lands

4. leaf-like covering of a flower

5. process whereby plants create glucose from sunlight, water and carbon dioxide

6. colorful part of flower which attracts pollinators

7. particle which contains cells which create pollen tube and sperm cells

9. plants are considered to be this because they can make their own food.

12. portion of male flower part that produces pollen grains

Name _____Date_____

Lesson 18   Practice Page 1

Please fill in the blank(s) for each statement below. Use your lesson to help you.

1. Members of the Kingdom Monera are considered to be _____ meaning
they have no nuclear membrane nor organized organelles.

2. Members of the Kingdom Monera also are _____ meaning they solely consist of being one cell.

3. The primary forms of reproduction in the Kingdom Monera is through _____ and
_____.

4. The phylum of Monera which includes the bacteria which causes disease is the Phylum
_____.

5. Members of this phylum (question 4) live off of _____ or live as parasites of
living things.

6. Viruses consist of a strand of _____ or _____ enclosed by a
_____ known as the _____.

7. Viruses are considered to be _____ _____ as they must live
within the cells of a host to survive.

8. Control of viruses is mainly through the use of _____ which stimulate the body's immune system to produce _____ to fight off the disease when encountered at a later
time.

9. Viruses work by entering a host cells and taking over _____ of the cell's DNA in an
effort to get the cell to make more viruses. The host cell then dies.

10. Viruses are classified as either being DNA or _____ viruses.

Name _____ Date_____

Lesson 18   Practice Page 2

Use the clues below to solve this crossword puzzle regarding the Kingdom Moneran.

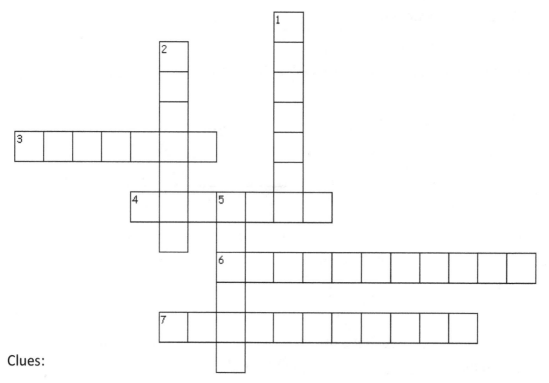

Clues:

Across

3. "living" things which cause disease

4. agent given to stimulate immune system to fight off virus should it ever arrive

6. have no cell membrane nor organized organelles

7. meaning only one-cell-big

Down

1. means of asexual reproduction utilized by monerans

2. kingdom of unicellular, prokaryotes

5. protein covering of a virus

Name _____Date_____

Lesson 19   Practice Page 1

Please fill in the blank(s) for each statement below. Use your lesson to help you.

1. Members of the Kingdom Protista are all _____ creatures meaning they are made up of organisms of only one cell.

2. They differ from members of the Kingdom Monera in that they have _____ or-ganelles and are therefore identified as being _____ as opposed to the Monerans which are prokaryotic.

3. While there are several phyla of the Kingdom Protista, these can be grouped into three main groups. These are the _____-like protists, the _____-like protists and the _____-like protists.

4. Members of the animal-like protists are grouped based upon _____.

5. _____ are the false-foot like appendages of Phylum Sarcodina.

6. _____ are the hair-like projections found in the Phylum _____.

7. Members of the Phylum _____ move about through the use of a flagellum which is a tail-like structure capable of propelling the organism through the water.

8. Members of the Phylum _____ are animal-like yet don't move about. Mem-bers include the _____.

9. The fungal-like protists include slime molds which move about on the _____ eat-ing up dead leaves.

10. The devastating potato famine of the 1800's was caused by a member of the _____ phylum.

11. The plant-like protists include various phyla of _____. They are categorized based upon the presence of various colors of _____ as well as how they store _____.

12. Chlorophyta are the _____ algae. Rhodophyta are the _____ algae.

13. Phaophyta are the _____ algae. Chrysophyta are the _____ algae.

14. Pyrrophyta have the ability to produce light through a process known as _____.

Lesson 19   Practice Page 2

Use the clues on the next page to solve this crossword puzzle regarding the Kingdom Protista.

Clues:

Across

2. algae capable of producing light

5. animal-like protists

7. fungal-like protists of forest floor

10. green algae

11. brown algae

14. hair-like projection allowing for movement

Down

1. living thing capable of producing light

3. red algae

4. false foot

6. golden brown algae

8. having membrane-bound organelles

9. tail-like structure of protozoans

12. plant-like protists

13. *Phytophthora spp.* caused this in Ireland in the mid-1800s

Name _____Date_____

Lesson 20   Practice Page 1

Please fill in the blank(s) for each statement below. Use your lesson to help you.

1. Fungi are _____meaning they live off of dead organisms. They are known as

_____.

2. Can be categorized according to the presence or lack of _____ in their hyphae.

Hyphae can be considered to be like _____ which grow through the substance being "eaten" by

the fungus.

3. Fungi can reproduce _____ and _____.

4. Fungi which belong to the Division Zygomycota have no _____ in their hyphae. The

common _____ is a member of this division.

5. Fungi which belong to the Division Basidiomycota have _____ in their hyphae. Fa-

miliar members include _____.

6. Fungi which belong to the Division Ascomycota have cross walls with _____and

reproduce through the use of _____or sacs which produce _____. Morels and

_____ are members of this division.

7. Fungi which belong to the Division Deuteromycota have _____ cycles

which are not yet fully understood. Athlete's foot fungus and the fungus used to create

_____ and flavor _____ belong in this division.

64

Name _____ Date_____

Lesson 20   Practice Page 2

Use the clues below to solve this crossword puzzle regarding the Kingdom Fungi

Clues:

Across

7. fungi can reproduce...

Down

1. Example of fungi in the Ascomycota division

2. this can be flavored using members of the Division Deuteromycota

3. mushrooms on your pizza are in this division of fungi

4. these fungi have no crosswalls in their hyphae

5. organisms which live off of dead organisms

6. "roots" of fungi

Name _____Date_____

Lesson 21   Practice Page 1

Please fill in the blank(s) for each statement below. Use your lesson to help you.

1. The study of how living things are made is known as _____ while how living things function, is known as _____.

2. At birth, a baby has over _____ bones of which several fuse together to reduce the number down to about _____ in an adult.

3. There are the _____ bones which consist of the bones making up the arms and legs, the _____ bones which are bones primarily of the skull and pelvis and the _____ bones which are the bones of the spinal column

4. Initially, the long bones were formed of _____ which over time is replaced by _____ and _____ containing compounds which make the bone hard. This hardening process is called _____.

5. On long bones, there are two _____, one found on each end and then one _____ which joins the two epiphyses.

6. Locations where ossification begins are known as _____.

7. The boundary where growth of the bone occurs is referred to as the _____or _____ of the bone.

8. The joints between of the long bones of the arms and legs are known as _____ joints.

9. If you look at the ends of long bones, you'll find a layer of _____cartilage.

10. To reduce friction between the ends of two adjoining bones, a fluid known as _____ keeps the bones gently sliding over each other

11. The tough membranes which bridge the joint space between the two bones is known as the

_____

12. Over time and with age, this slippery joint surface may become worn and result in pain and in-flammation. This is known as _____.

13. The joint between the femur and the pelvis at the hip is classified as a _____ joint.

14. The joint at the knee and elbow are classified as _____ joints as the mainly allow for bending in one direction.

15. The cells which make up bone are known as _____.

16. There are two kinds of osteocytes. The _____ are bone cells which have the responsibility of building new bone, like along the growth plate, while the _____ have the job of taking apart or tearing-down bone.

17. Bones are also very important when it comes to _____ of vital organs in the body. The flat bones of the skull protect the _____ while the irregularly-shaped bones of the spinal column protect the _____. The _____ also are very important for protecting the lungs, heart and upper abdominal organs.

18. The production of red blood cells takes place in the _____ of bone This tissue is known a hemopoietic tissue which literally means blood-making tissue. The marrow cavity is also a location for _____ in adults.

19. Bones also function as a storage site or depot for the element _____. Not only is calcium important for the _____ of our bones, calcium plays a major role in our

_____.

20. There are three types of muscle in the body. There is _____which is found in our internal organs like our stomach, intestines, bladder and respiratory tract. Then there is _____ which is specifically made for our heart. Finally, there is _____ which enables us to move.

21. Skeletal muscle is also referred to as _____ muscle.

22. There are components of skeletal muscle cells made up of what are known as _____ proteins or filaments. There are two types of contractile proteins: _____ and _____.

23. The attachment of muscles to bones is made through tough connective tissue material known as

_____.

24. Connective tissue structures which connect bones to bones are known as _____.

25. Muscles can only actively _____.

26. The end of the muscle which is nearer to the main trunk (chest and abdomen) of the body is

known as the _____ of the muscle. The opposite end of the muscle which is farther from the main trunk of the body is known as the _____ of the muscle.

27. Bending of a joint is referred to as _____ while straightening a joint is known as _____ the joint.

28. Inward movement is referred to as _____ and outward motion is referred to as _____.

29. Cells of the nervous systems are called _____ and consist of a central nucleus-containing area known as the _____. A single _____ is present with multiple _____.

30. An action potential (stimulus) travels into the _____ "end" and out through the _____"end" of a neuron.

31. Axons terminate on _____ (muscle cells) where _____ are released to activate muscle contraction. The most common neurotransmitter is _____ (ACH).

32. Neurons which carry messages out to muscles are known as _____ or _____ neurons.

33. Neurons which carry messages inward from sensory organs such as the skin or eyes are called _____ or _____ neurons.

34. Conduction of a stimulus is the result of _____ and _____ being pumped in and out of the neuron. The element _____ mediates this process.

35. _____ interferes with transmission of nerve impulses or neurotransmitter function which results in the desired control of pain and/or movement.

Name _____ Date_____

Lesson 21   Practice Page 2

Use the clues on the next page to solve this crossword puzzle regarding the skeletal, muscular and nervous systems.

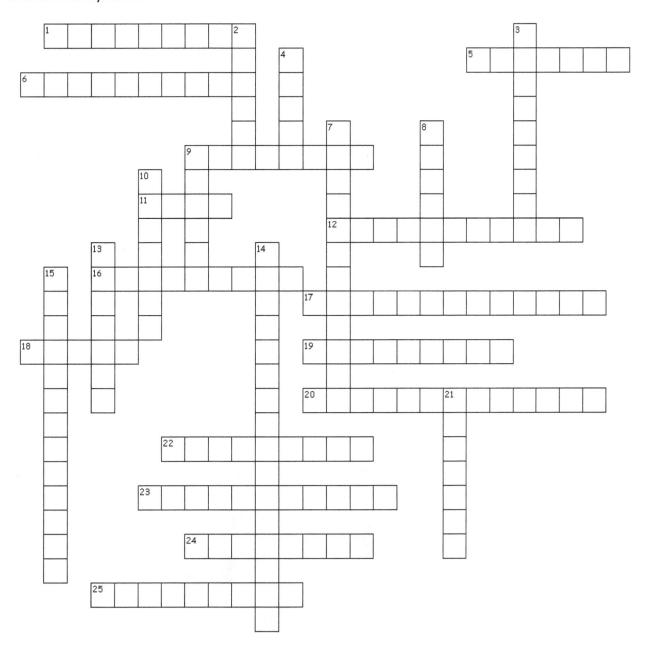

Clues:

Across

1. bones of arms and legs

5. making joint angle less

6. type of osteocyte which breaks down bone

9. type of muscle which moves bones

11. "end" of neuron where stimulus departs

12. type of osteocytes which builds bone

16. ends of long bones

17. hip joint

18. type of neuron which results in muscle action

19. part of bone between epiphyses

20. type of joint between long bones

22. making joint angle greater

23. contractile proteins of muscles

24. only intentional action of a muscle

25. bones of skull and pelvis

Down

2. type of joint between flat bones

3. "end" of neuron where stimulus leaves, multiple

4. elbow joint

7. location for red blood cell production

8. nerve cell

9. type of muscle of internal organs other than heart

10. element stored in bones

13. type of neuron which carries sensory information back to brain for interpretation

14. chemicals which move from end of nerve to muscle to trigger contraction

15. lubricant of joints

21. study of parts of body

Name _____ Date_____

Lesson 22   Practice Page 1

Please fill in the blank(s) for each statement below. Use your lesson to help you.

1. The primary function of the circulatory system is to deliver _____ and _____ to all cells of the body.

2. _____ are the smallest blood vessels, are only _____ cell layer in thickness and have windows or openings known as _____ to allow certain substances to pass.

3. The _____ of the fenestrations of capillaries varies throughout the body with some areas being very tightly controlled and other areas like in the _____ where whole blood cells freely move in and out of the capillaries.

4. Capillaries are "fed" by _____ which in turn receive blood from _____ which get blood from the largest artery of all which is the _____.

5. In the human heart there are four _____ . The two which receive blood from outside the heart are known as the _____ while the two which pump blood out of the heart are known as the _____.

6. Blood from the body that enters the heart is _____ in oxygen content and _____ in carbon dioxide. The chamber of the heart it enters is the _____ atrium.

7. From this chamber, upon contraction of the right atrium, blood moves down through the _____ valve into the _____ ventricle.

8. Upon contraction of the right ventricle, blood moves out of the heart destined for the _____ to get _____ and leave behind _____.

9. Blood returning from the lungs is _____ in oxygen and _____ in carbon dioxide. It enters the heart in the _____ atrium.

10. From the left atrium, blood moves downward through the _____ valve into the _____ ventricle.

11. Upon contraction, blood leaves the left ventricle passing through the _____ valve heading to the _____ through the largest artery known as the _____.

12. The right and left atrioventricular valves consist of flap-like structures known as the _____ of the valves. Preventing these structures from prolapsing upward into the atria above are tough string-like structures known as _____.

13. If a valve does not close properly which allows blood to "leak" in an unintended direction or is too "tight" (which is known as _____,) one can hear abnormal sounds known as _____.

14. Air moves down from the nose and mouth through the _____ which branches into _____ which branch further into _____.

15. The bronchioles end in small grape-like clusters of air sacs known as _____.

_____ wrap around the alveoli to pick up oxygen and leave behind _____.

16. In the body, capillaries returning from cells deliver blood into the smallest veins known as _____.

17. Unlike arteries, veins have _____ walls and have small _____ within them to prevent backflow of blood.

18. The watery portion of blood is known as _____.

19. Red blood cells are _____ while the scientific name for the group of white blood cells are the _____.

20. Red blood cells are produced in the _____ and have the primary job of _____ through the work of a special protein found within them known as _____.

21. The decrease in the functioning number of red blood cells is known as _____.

22. _____ function as part of our body's defense system against invading organisms.

23. _____ are white blood cells which create antibodies or _____ which work as _____ for later attack of invaders.

24. _____, _____ and _____ work to mediate inflamma-tion in the body.

25. Carbon monoxide causes problems in our body because it replaces _____ being carried by _____ within red blood cells.

Name _____ Date_____

Lesson 22   Practice Page 2

Use the clues below to solve this puzzle.

Across

3. tiny veins

6. smaller than an artery but larger than a capillary

7. flap portion of valve

8. most abundant leukocyte

9. chambers of heart which pump blood out of heart

10. location of red blood cell production

11. necessary substance for respiration

13. oxygen-carrying protein of red blood cells

17. low number of erythrocytes

20. red blood cells

21. watery portion of blood

Down

1. smallest blood vessel

2. "strings" of valve which prevents valvular prolapse

4. white blood cells which makes antibodies

5. windows of capillaries

12. "gate" within heart which allows for one direction flow

14. white blood cells

15. fuel of body carried by blood

16. largest artery of all

18. chambers of heart which welcome blood

19. pumper of blood

Name _____ Date_____

Lesson 23   Practice Page 1

Please fill in the blank(s) for each statement below. Use your lesson pages to help you.

1. The main function of the digestive system is to take _____ and convert it into glucose to be delivered to _____ by the _____ system.

2. A bite of food is known as a _____.

3. Saliva, which contains _____, is mixed with the bolus to begin the digestion process.

4. Food moves from the mouth to the stomach through the _____ by the action of _____muscles.

5. The stomach mixes the food with substances produced by the chief and parietal cells. The chief cells create _____ which reacts with hydrochloric acid from the _____ to produce pepsin, a _____ enzyme.

6. The "front door" of the stomach is the _____while the "back door" of the stomach known as _____.

7. The small intestine is divided into three segments: the _____ which is closely associated with the pancreas, the _____ where microvilli allow for great surface area for nutrient absorption and the _____.

8. The pancreas secretes the hormone _____ which regulates the ability of glucose to enter _____ of the body.

9. The _____ is known to have close to 500 specific jobs in the body which includes the _____ arriving from the small intestine as well as the manufacture of _____ and _____ (the storage form of glucose.)

10. At the junction of the ileum and the large intestine (colon) is the _____ or _____.

11. The primary role of the large intestine is _____ and the production and absorption of _____which is vital for blood _____.

Name _____Date_____

Lesson 23   Practice Page 2

Use the clues below to solve this crossword puzzle about the digestive system.

Across

2. this sphincter is the back door

6. process of taking food and converting to glucose

8. bite of food

9. pancreas secretes this hormone which allows glucose to enter cells

10. these are found in saliva to begin process of digestion

11. colon reabsorbs this

13. first section of small intestine with pancreas as a pal 14. allow for great surface area within small intestine

16. fancy name for spit

Down

1. enzymes which lyse proteins

2. cells of stomach which produce hydrochloric acid

3. cells of stomach which produce pepsinogen

4. tube from mouth to stomach, made of smooth muscle

5. longest section of small intestine

7. large intestine produces and absorbs this vitamin

12. blind tube portion of intestine

15. this sphincter is the front door

Name _____ Date_____

Lesson 24   Practice Page 1

Please fill in the blank(s) for each statement below. Use your lesson pages to help you.

1. The _____ and _____ systems of our body work to remove wastes created by cells as well as cell remnants from the body.

2. The renal system consists of our two _____ and blood vessels which carry "dirty" blood to and "clean" blood away from the _____.

3. The filtration unit of the kidney is the _____.

4. Waste products as well as water can leave through _____ in the walls of the arterioles within the glomerulus, however, _____, _____ and _____ are returned back to circulation and therefore conserved.

5. The kidneys also maintain _____ of the blood.

6. The kidneys produce hormones which control _____ and the creation of _____.

7. The presence of _____ in the urine indicates possible kidney damage or damage to other parts of the urinary system such as the _____ or _____.

8. Glucose in the _____ indicates it has exceeded the threshold capacity of the kidney.

_____, produced by the _____, functions to open "doors" of cell membranes to allow glucose to enter.

9. Lack of insulin can result in _____ levels of blood glucose, presence of glucose in the _____ and "starving" _____ all over the body.

10. Urine leaves the kidneys through the _____ which drain into the _____.

11. When full, the bladder is emptied to the exterior of the body through the _____.

Name _____Date_____

Lesson 24   Practice Page 2

Use the clues below to solve this crossword puzzle about the renal and urinary systems.

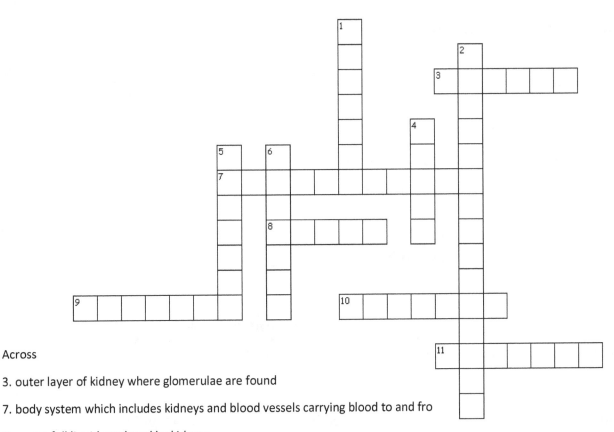

Across

3. outer layer of kidney where glomerulae are found

7. body system which includes kidneys and blood vessels carrying blood to and fro

8. waste-full liquid produced by kidneys

9. inner layer of kidney

10. presence of this in urine means too much is in blood

11. we have two and their job is to filter blood, regulate pH and blood pressure and RBC production

Down

1. these tubes carry urine from kidney to urinary bladder

2. these two elements are also "saved" by the kidneys and not flushed away

4. this gets conserved by kidneys so urine is not too watery

5. this tube empties bladder to exterior of body

6. this hormone produced by pancreas opens "doors" of cells to welcome glucose

Name _____Date_____

Please fill in the blank(s) for each statement below. Use your lesson pages to help you.

1. The _____ system is the body system which regulates the activities of many oth-

er body systems through the function of _____.

2. There are _____ types of glands: _____ which have ducts or tubes in which sub-

stances are exported and _____ glands which deposit substances directly into the

circulatory system.

3. Levels of hormones in the body are maintained using a _____ whereby low lev-

els trigger the endocrine gland to begin production of the hormone.

4. The _____ is known as the "master" gland as it produces hormones which affect

many organs in the body as well as other _____ glands which in turn affect body

organs.

5. The _____ pituitary gland produces thyroid stimulating hormone (TSH), adreno-

corticotropic hormone (ACTH), human growth hormone (HGH), prolactin and the gonadotropic hor-

mones.

6. TSH affects rate of _____ which is how fast or slow cells utilize _____.

7. ACTH signals the _____to produce anti-inflammatory substances and water con-

servation hormones.

8. HGH regulates _____ in the body.

9. Prolactin regulates development of _____ tissues as well as milk production in

women.

10.   The gonadotropic hormones in women stimulate the _____ to produce

_____ and then maintain pregnancy.

11.   The gonadotropic hormones in men stimulate the _____ to produce

_____ and _____.

12.   Hormones of the posterior lobe of the pituitary gland include _____ also known as _____ which regulates water conservation by the kidneys and oxytocin which stimulates the _____ to contract during birth and milk _____ while nursing.

13. The parathyroid glands control _____ levels in the body.

14. The _____ gland regulates daily sleep and wakefulness patterns.

15. The thymus, while only active in children, functions to produce _____ which function throughout life to defend against _____.

Lesson 25   Practice Page 2

Use the clues on the next page to solve this crossword puzzle about endocrine system.

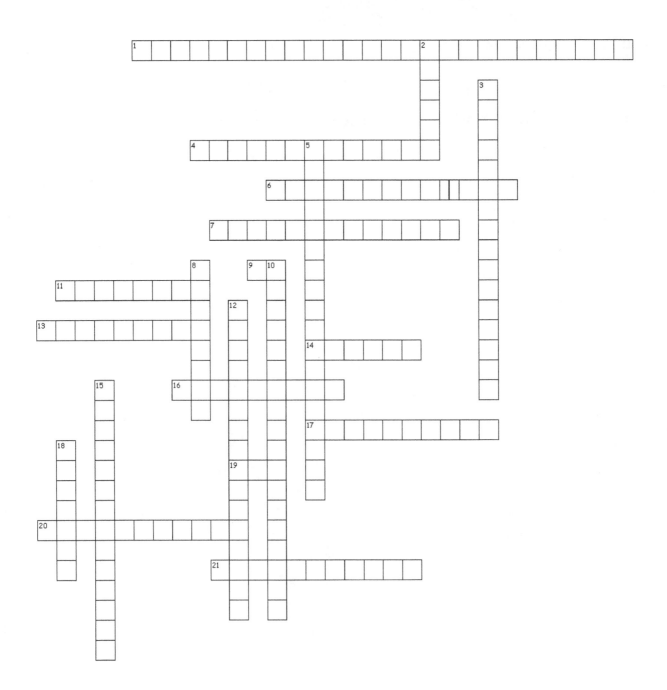

Across

1. this hormone stimulates development of follicles on the ovary in women

4. target organs of ACTH

6. the "master" gland

7. target organs of TSH

9. this hormone stimulates production of testosterone in men

11. this hormone stimulates the uterus to contract during labor as well as milk let-down during nursing

13. this body system utilizes hormones to regulate other body systems

14. target organs of gonadotropins in general

16. "back" side of pituitary gland

17. rate of glucose use by cells

19. this hormone stimulates production of sperm cells in men

20. this tiny gland regulates sleep and wakefulness patterns

21. produced by the posterior pituitary gland to regulate water levels in the body

Down

2. this childhood gland produces killer-T cells for lifelong defense strategies

3. adrenal cortex produces these agents to reduce pain, swelling and redness

5. this hormone stimulates ovulation in women

8. "front" side of pituitary gland

10. HGH which regulates growth is short for _____

12. this mechanism is based upon low levels triggering action upon producing glands

15. these glands have tubes or ducts to delivery products to body

18. levels of this important element are controlled by the parathyroid glands

Name _____Date_____

Lesson 26   Practice Page 1

Please fill in the blank(s) for each statement below. Use your lesson pages to help you.

1. Sperm and ova are _____ cells having under gone meiosis. Sperm are formed in the

_____ and ova form in the _____.

2. The testis are located outside the body in the _____. The ovaries are located in the

_____.

3. Sperm leave the testis through the _____ and continue through the

vas deferens to the _____.

4. Accessory sex glands which include the seminal vesicles, Cowper's gland and prostate gland pro-

vide _____ and _____ substances to the sperm to create _____.

5. The _____ allows the man to deposit semen into the woman's _____.

6. The cervix functions like a _____ to open at specific times in response to _____ to

allow sperm to enter or be closed during pregnancy.

7. The release of the ovum is known as _____. The ovum is gathered by the

_____ and moved down the fallopian tube where _____

usually takes place.

8. The embryo moves down to the _____ to implant and continue development.

9. Blood never mixes directly between the _____ and developing _____. Nu-

trients and wastes along with oxygen and carbon dioxide readily move across _____

membranes of the _____.

10. Pregnancy in humans is _____ months.

11. _____ from the posterior pituitary gland stimulates the uterus to begin

contractions while decreasing _____ levels allows the cervix to dilate.

12. Should the baby be too large to deliver, a _____ may be performed to

surgically remove the baby from the uterus.

Name _____Date_____

Lesson 26   Practice Page 2

Use the clues below to solve this crossword puzzle.

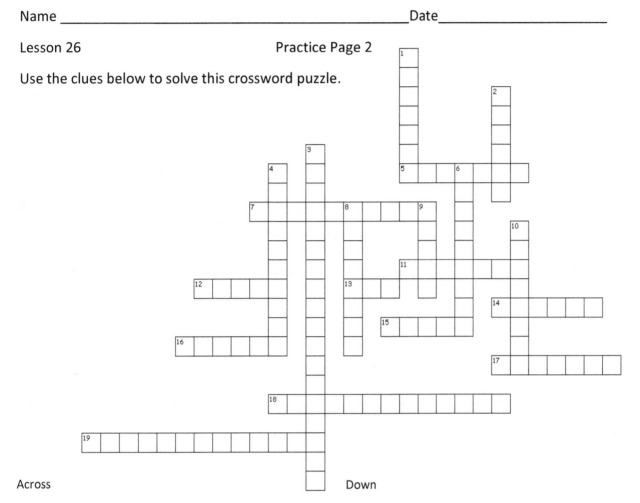

Across

5. process of going from 2N to 1N

7. length of time of pregnancy

11. tube leading from vas deferens to exterior of body

12. location for ovum development

13. female gametes

14. "door" of female reproductive system

15. liquid carrying sperm cells

16. location where embryo implants and grows throughout pregnancy

17. condition of having one-half the total number of chromosomes 1N

18. usual location of fertilization

19. process where gametes join to create new 2N individual

Down

1. location of testis outside body

2. location for sperm and testosterone production

3. includes seminal vesicles, Cowper's gland and prostate gland

4. tube leading from testis to vas deferens

6. process where follicle ruptures to release ovum

8. hormone which causes uterine contractions

9. male gametes

10. location where blood vessels from baby come close to blood vessels of mother but never actually join

Name _____Date_____

Lesson 27   Practice Page 1

Please fill in the blank(s) for each statement below. Use your lesson pages to help you.

1. _____, _____, _____,

_____, and _____ are our five means of sensing what it hap-

pening in our environment.

2. The clear outer covering of our eyes is called the _____ which transitions to the

_____ which is the white portion on the perimeter of the eye.

3. Tears from the _____ continually bathe the cornea to maintain hydration and wash away

_____.

4. The colored portion within the eye is the _____ and creates the circular _____

which adjusts the amount of _____ which can enter the eye.

5. The fluid in front of the lens of the eye is the _____ humor while the fluid behind the

lens, which is much thicker, is the _____ humor.

6. Excess fluid within the eye is _____.

7. Tiny muscles encircling the lens allow it to change _____ which in turn allows images to

be focused on the _____.

8. The _____ is the multilayered surface at the back of the eye which has specialized cells

capable of converting light into nerve stimulus.

9. Nerve endings in the retina gather into the _____ which transmits messages to the

_____.

10. The ears have the capability of converting _____ energy into _____which in

turn stimulate nerve endings to allow us to hear.

11. Sound moves into the auditory canal and strikes the _____ membrane which causes the

three tiny bones of the middle ear to vibrate. These vibrations stimulate nerve endings within the

fluid-filled _____ which then send signals by the _____ nerve to the brain.

12. The semicircular canals within the ear work to maintain _____.

13. In our nose there are nerve endings which are capable of taking _____ stimuli created by contact with tiny molecules of substances floating in the air and converting it to nerve impulses.

14. Stimuli within the nose travels to the brain by the _____ nerve.

15. The main organ of taste sensation is the _____ .

_____ on the surface of the tongue house taste buds which consist of taste receptor cells.

16. The primary organ of touch is our _____. When compared to all other organs in our body, the skin is the _____ organ and has the most _____ (weight).

17. The skin consists of _____ layers: the _____ which creates squamous cells that form the surface to the skin and the deeper _____ which houses blood vessels, nerves, glands and fatty tissue.

18. Touch sensations travel by _____ neurons to the spinal cord and brain.

19. The skin can be divided into regions of sensation. These regions are known as _____.

Name _____Date_____

Use the clues below to solve this crossword puzzle.

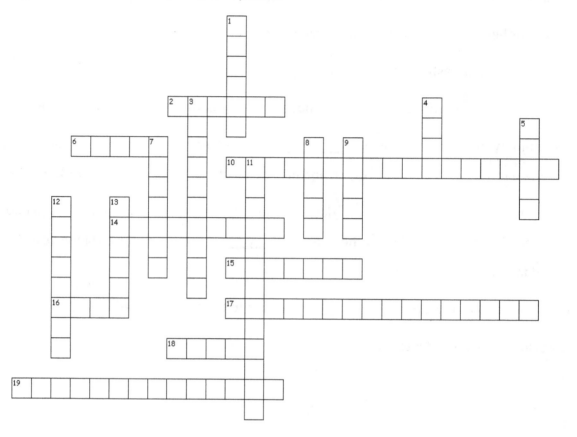

Across

2. clear covering of the eye

6. largest organ of body enables this sensation

10. three tiny bones of middle ear

14. outer layer of skin with flattened cells

15. fluid-filled snail-shaped structure of inner ear with nerve endings inside

16. curved "onion" which focuses light on retina

17. ear drum

18. sense of flavor gathered by the tongue

19. cranial nerve of scent

Down

1. light sensitive cell layer in rear of eye

3. cranial nerve which connects eyes to brain

4. colorful circle in the eye

5. "job" of the nose

7. three tiny bones work together to enable this sensation

8. hole created by the iris to adjust light entry

9. our eyes enable this ability

11. cranial nerve of hearing

12. clusters of taste buds on surface of tongue

13. deeper layer of skin with blood vessels, nerves, glands and fatty tissues

Name _____Date_____

Lesson 28   Practice Page 1

Please fill in the blank(s) for each statement below. Use your lesson pages to help you.

1. _____ is the study of the relationships between living things and their environment.

2. _____ is defined as being everything, both living and nonliving, which are found in the location where a living organism lives.

3. The _____ is the segment of the earth where life exists.

4. A _____ is defined as an area or region of the earth where similar organisms thrive and others do not.

5. The _____ biome is found at the north and south poles of the earth. The average temperature range is -40 C to -4 C (-40 F to 25 F) with less than 5 inches of precipitation in a year.

6. The temperature range of the _____ is -26 C to 4 C (-28 F to 39 F) and precipitation is less than 10 inches per year. _____ and a few mammals are found in this biome.

7. The _____ biome is characterized by evergreen, cone-bearing trees and has a temperature range of this biome is -10 C to 14 C (14 F to 57 F). The average yearly precipitation is between 12 and 30 inches.

8. The _____biome is characterized by the predominance of trees which are lose their leaves during the cooler seasons of the year. The average temperature of this biome is 6-28 C (42 F to 82 F) and annual precipitation is 30-50 inches.

9. At the same latitude of the deciduous forest biome is the _____ biome which experiences temperatures slightly cooler than the deciduous forest biomes and receive only 10-30 inches of precipitation in a year. Abundant grasses and few trees are present in this biome.

10. _____ biomes are found in regions which are warmer on average than grasslands yet receive less precipitation. The temperature range in desert biomes are 24-34 C (75 F to 94 F) with precipitation being less than 10 inches per year.

11. The rain forest biome can be divided into the _____ rain forest and the

_____ rain forest. The tropical rain forest is found all along the equator where the average temperature is 25-27 C (77 F to 81 F). The temperate rain forest is found along the western coast of North America where temperatures range from 10-20 C (50 F to 60 F).

12. The _____ biome consists of the earth's oceans and seas.

13. The _____ biome consists of lakes, rivers and streams.

_____ are where rivers of freshwater mix with the salt water of a sea or ocean.

14. The organisms which are capable of capturing energy from the sun and converting it into useable food for itself (and later, others) are known as _____. Autotrophs are also called _____.

15. Heterotrophs "harvest" energy captured from _____. These organisms are also known as _____.

16. The consumers which first eats a producer (plant) is known as a _____ consumer. Organisms which eat a primary consumer are called _____ consumers.

17. Consumers which have plants as their sole source of food are known as _____.

18. Consumers which eat both plants and animals (producers and consumers) are known as _____.

19. _____ are those organisms which consume dead organisms

20. With each step through an ecosystem away from the producer level, the amount of available energy is _____.

21. _____ cycles demonstrate how chemical elements move round-and-round through living (bio-) and then non-living portions (geo-) of the environment.

Name _____ Date_____

Lesson 28   Practice Page 2

Use the clues below to solve this crossword puzzle.

Across

6. eater of road kill

11. biome which receives most rainfall

13. biome characterized by abundant grasses and few trees

14. these are the trees found in the coniferous forest biome

17. you'll only find very short plants and a few mammals living here

18. those who consume autotrophs

19. eater of plants

Down

1. diagram showing how elements move about through the living and non-living segments of the earth

2. everything, both living and non-living, where an organism lives

3. diagram showing how things eat one another

4. trees which lose their leaves in the cool seasons of the year

5. eater of both plants and animals

7. study of relationships of living things and their environment

8. coldest biome of the earth

9. one who consumes producer

10. another name for autotroph

12. eater of animals

15. literally means self-feeder

16. location where some living things thrive and others do not

CPSIA information can be obtained
at www.ICGtesting.com
Printed in the USA
FSHW020201300321
79902FS